SOCIAL ENVIRONMENT AND MORAL PROGRESS

BY

ALFRED RUSSEL WALLACE

British Library Cataloguing-in-Publication Data
A catalogue record for this book is available from the
British Library

Alfred Russel Wallace

Alfred Russel Wallace was born on 8th January 1823 in the village of Llanbadoc, in Monmouthshire, Wales.

At the age of five, Wallace's family moved to Hertford where he later enrolled at Hertford Grammar School. He was educated there until financial difficulties forced his family to withdraw him in 1836. He then boarded with his older brother John before becoming an apprentice to his eldest brother, William, a surveyor. He worked for William for six years until the business declined due to difficult economic conditions.

After a brief period of unemployment, he was hired as a master at the Collegiate School in Leicester to teach drawing, map-making, and surveying. During this time he met the entomologist Henry Bates who inspired Wallace to begin collecting insects. He and bates continued exchanging letters after Wallace left teaching to pursue his surveying career. They corresponded on prominent works of the time such as Charles Darwin's *The Voyage of the Beagle* (1839) and Robert Chamber's *Vestiges of the Natural History of Creation* (1844).

Wallace was inspired by the travelling naturalists of the day and decided to begin his exploration career collecting specimens in the Amazon rainforest. He explored the Rio Negra for four years, making notes on the peoples and languages he encountered as well as the geography, flora, and fauna. On his return voyage his ship, Helen, caught fire and he and the crew were stranded for ten days before being picked up by the Jordeson, a brig travelling from Cuba to London. All of his specimens aboard Helen had been lost.

After a brief stay in England he embarked on a journey to the Malay Archipelago (now Singapore, Malaysia, and Indonesia). During this eight year period he collected more than 126,000 specimens, several thousand of which represented new species to science. While travelling, Wallace refined his thoughts about evolution and in 1858 he outlined his theory of natural selection in an article he sent to Charles Darwin. This was published in the same year along with Darwin's own theory. Wallace eventually published an account of his travels *The Malay Archipelago* in 1869, and it became one of the most popular books of scientific exploration in the 19th century.

Upon his return to England, in 1862, Wallace became a staunch defender of Darwin's landmark work *On the Origin of Species* (1859). He wrote responses to those critical of the theory of natural selection, including 'Remarks on the Rev. S. Haughton's Paper on the Bee's Cell, And on the Origin of Species' (1863) and 'Creation by Law' (1867). The former of these was particularly pleasing to Darwin. Wallace also published important papers such as 'The Origin of Human Races and the Antiquity of Man Deduced from the Theory of 'Natural Selection" (1864) and books, including the much cited *Darwinism* (1889).

Wallace made a huge contribution to the natural sciences and he will continue to be remembered as one of the key figures in the development of evolutionary theory.

Wallace died on 7th November 1913 at the age of 90. He is buried in a small cemetery at Broadstone, Dorset, England.

CONTENTS

SOCIAL ENVIRONMENT AND MORAL PROGRESS

PART I.--HISTORICAL

CHAPTER I: INTRODUCTORY

Before entering on the question of the relation of morality to our existing social environment, it will be advisable to inquire what we mean by moral progress, and what evidence there is that any such progress has occurred in recent times, or even within the period of well-established history.

By morals we mean right conduct, not only in our immediate social relations, but also in our dealings with our fellow-citizens and with the whole human race. It is based upon the possession of clear ideals as to what actions are right and what are wrong and the determination of our conduct by a constant reference to those ideals.

The belief was once prevalent, and is still held by many persons, that a knowledge of right and wrong is inherent or instinctive in everyone, and that the immoral person may be justly punished for such wrongdoing as he commits. But that this cannot be wholly, if at all, true is shown by the fact that in different societies and at different periods the standard of right and wrong changes considerably. That which at one time and place is held to be right and proper is, at another time or place, considered to be not only wrong, but one of the greatest of crimes. The most striking example of this change of opinion is that as to slavery, which was held to be quite justifiable by the most highly civilized people of antiquity, and hardly less so by ourselves within the memory of persons still living. The owners of sugar estates in Jamaica cultivated by slaves were not stigmatized as immoral by their relatives in England or by the public at large; and it was the horror excited by the slave-trade in Africa, and in the "middle passage" on the slave ships, rather than by the slavery itself, that so excited public opinion as to lead to the abolition first of the one and then of the other.

We are obliged to conclude, therefore, that what is commonly termed morality is not wholly due to any inherent perception of what is right or wrong conduct, but that it is to some extent and often very largely a matter of convention, varying at different times and places in accordance with the degree and kind of social development which has been attained often under different and even divergent conditions of existence. The actual morality of a community is largely a product of the environment, but it is local and temporary, not permanently affecting the character.

To bring together the evidence in support of this view, to distinguish between what is permanent and inherited and what is superficial and not inherited, and to trace out some of the consequences as regards what we term "morality" is the purpose of the present volume.

CHAPTER II: MORALITY AS BASED UPON CHARACTER

Though much of what we term morality has no absolute sanction in human nature, yet it is to some extent, and perhaps very largely, based upon it. It will be well, therefore, to consider briefly the nature and probable origin of what we term "character"--in individuals, in societies, and especially in those more ancient and more fundamental divisions of mankind which we term "races."

Character may be defined as the aggregate of mental faculties and emotions which constitute personal or national individuality. It is very strongly hereditary, yet it is probably subject to more inherent variation than is the form and structure of the body. The combinations of its constituent elements are so numerous as, in common language, to be termed infinite; and this gives to each person a very distinct individuality, as manifested in speech, in emotional expression, and in action.

The mental faculties which go to form the "character" of each man or woman are very numerous, a large proportion of them being such as are required for the preservation of the individual and of the race, while others are pre-eminently social or ethical. These latter, which impel us to truth, to justice, and to benevolence, when in due proportion to all the other mental faculties, go to form what we distinguish as a good or moral character, and will in most cases result in actions which meet with the general approval of that section of society in which we live; and this approval reacts upon the character so that it often appears to be better than it really is.

So great is the effect of this approval of our fellows that it sometimes leads to behavior quite different from what it would be if this approval were absent. This is especially the case when the approval leads to wealth or positions of dignity or advantage. Occasionally, in cases of this kind, the individual cannot resist his natural impulses, and then acts so as to show his underlying real character. We term such persons hypocrites for making us believe that they were inherently good, instead of being so in appearance only when the good action was profitable to them. Hence in a highly complex state of civilization it becomes exceedingly difficult correctly to appraise characters as moral or immoral, good or bad; while there is no such difficulty as regards the intellectual and emotional aspects of character, which are less influenced by the general environment, and which there is less temptation to conceal.

All the evidence we possess tends to show that although the actions of most individuals are to a considerable extent determined by their social environment, that does not imply any alteration in their character. Everyone's experience of life, and especially the example of his friends and associates, leads him to repress his passions, regulate his emotions, and in general to use his judgment before acting, so as to secure the esteem of his fellows and greater happiness for himself; and these restraints, becoming habitual, may often give the appearance of an actual change of character till some great temptation or violent passion overcomes the usual restraint and exhibits the real nature, which is usually dormant.

Now it is this inherent and unchangeable character itself that tends to be transmitted to offspring, and this being the case, there can be no progressive improvement in character without some selective agency tending to such improvement. By means of a general discussion of the nature and origin of "Character," I have elsewhere shown that there is no proof of any real advance in it during the whole historical period.* [*See *Character and Life*, edited by P. L. Parker, p-31. (Williams and Norgate; November, 1912.)] I show later on what the required selective agency is, and how it will come into action automatically when, and not until, our social system is so reformed as to afford suitable conditions. (*See* Chapter XVI.)

CHAPTER III: PERMANENCE OF CHARACTER

I will now call attention to a few of the facts which lead to the conclusion as to the stationary condition of general character from the earliest periods of human history, and presumably from the dawn of civilization. In the earliest records which have come down to us from the past we find ample indications that general ethical conceptions, the accepted standard of morality, and the conduct resulting from these, were in no degree inferior to those which prevail to-day, though in some respects they differed from ours.

As examples of great moral teachers in very early times we have Socrates and Plato, about 400 B.C.; Confucius and Buddha, one or two centuries earlier; Homer, earlier still; the great Indian Epic, the Maha-Bharata, about 1500 B.C. All these afford indications of intellectual and moral character quite equal to our own; while their lower manifestations, as shown by their wars and love of gambling, were no worse than corresponding immoralities today.

In the beautiful translation by the late Mr. Romesh Dutt, of such portions of the Maha-Bharata as are best fitted to give English readers a proper conception of the whole work, there is a striking episode entitled "Woman's Love," in which the heroine, a princess, by repeated petitions and reasonings persuades Yama, the god of death, to give back her husband's spirit to the body. It is described in the following verses:

"And the sable King was vanquished, and he turned on her again,
And his words fell on Savitri like the cooling summer rain:
'Noble woman, speak thy wishes, name thy boon and purpose high,
What the pious mortal asketh gods in heaven may not deny!'
"'Thou hast,' so Savitri answered, 'granted father's realm and might,
To his vain sightless eyeballs hath restored the blessed light;
Grant him that the line of monarchs may not all untimely end,
That his kingdom to Satyavan and Savitri's sons descend!'
"'Have thy wishes,' answered Yama; 'thy good lord shall live again,
He shall live to be a father, and your children, too, shall reign;
For a woman's troth endureth longer than the fleeting breath,
And a woman's love abideth higher than the doom of death.'"

And when at the end of the epic, the kings and warriors welcome each other in the spirit world, we find the following noble conception of the qualities and actions which give them a place there:

"These and other mighty warriors, in the earthly battles slain,
By their valor and their virtue walk the bright ethereal plain!
They have lost their mortal bodies,
    crossed the radiant gate of heaven,

For to win celestial mansions unto mortals it is given!
Let them strive by kindly action, gentle speech, endurance long,
Brighter life and holier future unto sons of men belong!"

Mr. Dutt informs us that he has not only reproduced, as nearly as possible, the metre of the original, but has aimed at giving us a literal translation. No one can read his beautiful rendering without feeling that the people it describes were our intellectual and moral equals.

The wonderful collection of hymns known as the Vedas is a vast system of religious teaching as pure and lofty as those of the finest portions of the Hebrew scriptures. A few examples from the translation by Sir Monier Monier-Williams will show that its various writers were fully our equals in their conceptions of the universe, and of the Deity, expressed in the finest poetic language. The following is a portion of a hymn to "The Investing Sky":

"The mighty Varuna, who rules above, looks down
Upon these worlds, his kingdom, as if close at hand.
When men imagine they do aught by stealth, he knows it.
No one can stand or walk, or softly glide along
Or hide in darkness, or lurk in secret cell
But Varuna detects him and his movements spies.

* * * *

This boundless earth is his,

His the vast sky, whose depth no mortal e'er can fathom.
Both oceans find a place within his body, yet
In the small pool he lies contained; whoe'er should flee
Far, far beyond the sky would not escape the grasp
Of Varuna, the king. His messengers descend
Countless from his abode--for ever traversing
This world, and scanning with a thousand eyes its inmates.
Whate'er exists within this earth, and all within the sky,
Yea, all that is beyond King Varuna perceives.
May thy destroying snares cast sevenfold round the wicked,
Entangle liars, but the truthful spare, O King."

The following passage from a "Hymn to Death," shows a perfect confidence in that persistence of the human personality after death, which is still a matter of doubt and discussion today:

"To Yama, mighty king, he gifts and homage paid.
He was the first of men that died, the first to brave
Death's rapid rushing stream, the first to point the road
To heaven, and welcome others to that bright abode.
No power can rob us of the home thus won by thee.
O king, we come; the born must die, must tread the path
That thou hast trod--the path by which each of men,
In long succession, and our fathers, too, have passed.
Soul of the dead! depart; fear not to take the road--
The ancient road--by which thy ancestors have gone;

Ascend to meet the god--to meet thy happy fathers,
Who dwell in bliss with him.
Return unto thy home, O soul! Thy sin and shame
 Leave thou behind on earth; assume a shining form--
Thy ancient shape--refined and from all taint set free."

In this we find many of the essential teachings of the
most advanced religious thinkers--the immediate entrance
to a higher life, the recognition of friends, the persistence
of the human form, and the shining raiment, typical of the
loss of earthly taint.

But besides these special deities, we find also the
recognition of the one supreme God, as in the following
hymn:

"What god shall we adore with sacrifice?
Him let us praise, the golden child that rose
In the beginning, who was born the Lord--
The one sole lord of all that is--who made
The earth, and formed the sky, who giveth life,
Who giveth strength, whose bidding gods revere,
Whose hiding place is immortality,
Whose shadow, death; who by his might is king
Of all the breathing, sleeping, waking world--
Who governs men and beasts; whose majesty
These snowy hills, this ocean with its rivers,
Declare; of whom these spreading regions form

The arms by which the firmament is strong,
Earth firmly planted, and the highest heavens
 Supported, and the clouds that fill the air
Distributed and measured out; to whom
Both earth and heaven, established by his will,
Look up with trembling mind; in whom revealed
The rising sun shines forth above the world."

If we make allowance for the very limited knowledge of Nature at this early period, we must admit that the mind which conceived and expressed in appropriate language such ideas as are everywhere apparent in these Vedic hymns could not have been in any way inferior to those of the best of our religious teachers and poets--to our Miltons and our Tennysons.

CHAPTER IV: PERMANENCE OF HIGH INTELLECT

Accompanying this fine literature and moral teaching in Ancient India was a civilization equal to that of early classical races, in grand temples, forts and palaces, weapons and implements, jewelry and exquisite fabrics. Their architecture was highly decorative and peculiar, and has continued to quite recent times. Owing perhaps to the tropical or sub-tropical climate, with marked wet and dry seasons, the oldest buildings that have survived, even as ruins, are less ancient than those of Greece or Rome--but those corresponding in age to the period of our Gothic cathedrals are immensely numerous, and show an originality of design, a wealth of ornament, and a perfection of workmanship equal to those of any other buildings in the world.

Two other great civilizations of which we have authentic records are those of Egypt and Mesopotamia, both of which appear to have been much older than those of India or Greece. But whereas Egypt has left us the most continuous series of tombs, temples, and palaces in the world, abundant works of art in statues and sculptures, together with characteristic reliefs and wall paintings, showing the whole public and domestic life of the people, Mesopotamia is represented only by vast masses of ruins on the sites of the ancient cities of Nineveh and Babylon, from which have been disinterred many fine statues and reliefs, exhibiting a very distinct style of art. For more than 2,000 years the history and remains of this once greatest of civilizations was absolutely unknown, except by a few doubtful facts and names in Greek and Hebrew writings. But during the latter half of the nineteenth century a band of explorers and students, such as Layard and Rawlinson, made known, first the works of art, and, latterly, an enormous quantity of small bricks and stone slabs, thickly covered with a peculiar kind of writing known as the cuneiform inscriptions, which, after an enormous amount of labor, have at length been translated. Whole libraries of these brick-books have been discovered, and as the reading and translating goes on we obtain a knowledge of the history, laws, customs, and daily life of this ancient people almost equal to that we now possess of the ancient Indians and Egyptians.

For our present purpose, however, Egyptian civilization is the most important, because it presents us with the most definite proof of the attainment of a high degree of what is specially scientific attainment at the very dawn of historical knowledge. This is well exhibited by that most wonderful work of constructive art--the Great Pyramid of Gizeh--which, though not quite the earliest, is the largest and most remarkable of about seventy pyramids in various parts of

Egypt, and has been more thoroughly explored and studied, both as to its proportions, construction and uses, than any of the others.

This pyramid is known historically to have been built by the order of King Cheops (or Khufu), and the date of its design and erection can be pretty accurately fixed as about 3700 B.C., or nearly 2,000 years earlier than that of the civilization depicted in the Indian and Greek epics. The internal structure of this pyramid is its most interesting feature, because it shows clearly that it was designed to be not only the tomb of the king who built it, but also a true astronomical observatory during his life. This has been denied by some modern historians. In Harmsworth's *History of the World* () it is said: "For the pyramids are nothing but tombs. They have no astronomical meaning or intention whatever." And then, after referring to the ideas of Piazzi Smyth and others as "vain imaginings," it is added: "There is nothing marvellous about these great tombs except their size and the accuracy of their building." An almost exactly similar statement is made in the great *Historian's History of the World* and in "Chambers's Encyclopædia."

If the writers of these histories had read Mr. R. A. Proctor's book, *The Great Pyramid: Observatory, Tomb, and Temple*, they would have known that this statement is entirely erroneous. The size, shape, and angles of the internal passages have been described and measured by many competent students, among the most careful and exact of whom was Piazzi Smyth, then Astronomer Royal of Scotland. It is true he had many "vain imaginings," but his measurements were among the most trustworthy. The "pyramid religion," which he helped to establish by a series of "coincidences" in the dimensions of various parts of the pyramid with astronomical dimensions, of which the pyramid builders could have had no knowledge whatever (such as the distance of the sun, the precession of the equinoxes, etc.), was no doubt a "vain imagining," but he frankly claimed it as a divine inspiration. All these are rejected by Mr. Proctor, who clearly explains the purpose of the greater part of the internal structure as only an experienced practical astronomer could do. I will now state as briefly as possible what are the well-established facts as well as the conclusions at which Mr. Proctor arrives.

The Great Pyramid and the two smaller ones near it, forming the pyramids of Gizeh, are placed on a small rocky plateau near the apex of the delta of the Nile. The largest of these is situated so that its northern face rises from the very edge of this plateau. The reason of this seems to have been that the builders wished to place it as nearly as possible on the 30th parallel of latitude. It is really about a mile and a third south of that parallel, and it is shown that such an error is a small one for that early period, and would matter but very little for the purpose required. The next feature is that it is truly oriented; that is, the four sides run north and south, east and west. It is also a true square, the four sides being of equal length, and the four corners are on a truly level plane.

The first thing the builders had to do was to get a true meridian line, and they could have done this in two ways--by observations of the sun or of the polestar, the latter being much the more accurate, though more laborious and costly. At the time the pyramid was built the polestar was Alpha Draconis, which was farther from the pole than our polestar and revolved around the true pole in a circle of 7° 24' in diameter. In order to observe the direction of this star at its lowest point, the builders excavated in the solid rock a tunnel about 4 feet in diameter, so as to keep this star visible each day at the lowest point of its circuit. This tunnel extended 350 feet through the rock to a point nearly under the centre of the pyramid, where, by a small vertical boring, a plumb-line could have been dropped so as to obtain the exact line of the meridian on the surface, and afterwards on each successive step of the pyramid as it was built up. While the building went on the sloping tunnel was continued backwards to its northern face, and a tunnel ascending to the south was formed of the same size and making the same angle with the horizon. This had puzzled all previous explorers of the pyramid till Mr. Proctor showed that, by stopping up the downward passage at the angle and filling the hollow with water, the polestar could be observed by reflexion and thus give the exact direction of the meridian on the upper surface of the pyramid with extreme accuracy, as it was built up slowly year by year.

But at a distance of 127 feet a new feature appears. The ascending tunnel is changed into what is called the Great Gallery, which, while continuing exactly the same floor line as the tunnel, is suddenly raised to a height of 28 feet, with a width of 7 feet on the floor and 3 1/2 feet at the top. Along each side there is a ledge or seat, 20 inches broad and 21 inches high. The sides do not slope inwards, but are formed of seven courses of stone, each one overlapping the one below by about 3 inches. The whole of this gallery, or inclined corridor, is formed of limestone, beautifully smooth, or even polished. The length of this gallery is 156 feet, and its floor terminated at the platform of the pyramid, upon the central line from east to west, when it had reached two-thirds of its total height. This is on the level of the King's Chamber; and it was probably only after the king was dead and his body embalmed and placed in his sarcophagus that the pyramid was completed, the openings of the passages carefully closed up, and the whole exterior covered with a smooth casing of stone, very small portions of which now remain. There are two other features of this gallery which have puzzled the merely antiquarian explorers. These are square holes cut in the sloping benches close to the side walls, and about 5 1/2 feet apart, there being eighteen on each side exactly opposite each other. On each side of the gallery, about half-way up, is a longitudinal groove, which would serve to carry transverse screens which could be slid up or down and easily wedged in position in order to mark exactly the central line, like the cross hairs in an astronomical telescope. The holes on the benches would serve to carry cross seats on which the observer could be firmly and comfortably seated while observing a transit of sun, star or planet.

Being open to the south, the Great Gallery would give a magnificent view of the southern sky, and enable observers to determine the altitudes and azimuths of many stars, and of the superior planets Mars, Jupiter, and Saturn. The star Alpha Centauri, which was at that period of the first magnitude, though now much diminished in brightness, would, when crossing the meridian, have been situated about the centre of the field of view as seen from this remarkable feature of the pyramid which, Mr. Proctor considers, was the finest transit-instrument ever constructed for naked-eye observations. Tycho Brahé, with his celebrated Quadrant at Uranienburg, did not attain such a degree of accuracy as did these Eastern astronomers nearly 6,000 years ago. One great superiority of the subterranean observatory over any open-air observations that can be made without telescopes is, that by closing up the end, except for the small aperture required to see the object, the brighter stars could be well observed in the daytime.

When we remember that the Great Pyramid covers 13 1/2 acres of ground, that it is truly square and on a truly horizontal base, that each side is accurately directed to a point of the compass, that the angle of its slope is such that the area of each of the four triangular faces is equal to that of a square whose sides are equal to the height of the pyramid; and, further, that the slope of the long descending tunnel is precisely such as to point accurately to the polestar of the epoch at the lowest part of its circuit round the true pole; and, lastly, that all this could only be done, as accurately as it has been done, by the system of subterranean tunnels and galleries that actually exists, while almost all the details of their construction are shown to be adapted for astronomical observations of the nature required, the conclusion becomes irresistible that they were designed and used for such observations, and that by no other means could the same amount of accuracy have been attained.

I have given a rather full account of what the Pyramid builders really did, because it forms a very important part of the argument I am developing as to the stationary condition of the human intellect during the historical period.

The great majority of educated persons hold the opinion that our wonderful discoveries and inventions in every department of art and science prove that we are really more intellectual and wiser than the men of past ages--that our mental faculties have increased in power. But this idea is totally unfounded. We are the inheritors of the accumulated knowledge of all the ages; and it is quite possible, and even probable, that the earliest steps taken in the accumulation of this vast mental treasury required even more thought and a higher intellectual power than any of those taken in our own era.

We can perhaps best understand this by supposing any one of our great men of science to have been born and educated in one of the earliest of the civilizations. If Newton had been born in Egypt in the era of the Pyramid builders, when there were no such sciences as mathematics, perhaps even no decimal notation which makes arithmetic so easy to us, he could probably have done nothing more than they have actually done. In building up the sciences each of the early steps was the work of a genius. But now that there have been nearly a hundred centuries of discovery and specialization by thousands or even millions of workers, that by means of writing and of the printing-press every discovery is quickly made known, and that ever larger and larger numbers devote their lives to study, the rate of progress becomes quicker and quicker, till the total result is amazingly great. But that does not prove any superiority of the later over the earlier discoveries. There is, therefore, no proof of continuously increasing intellectual power.

But we have now evidence of another kind, which adds to the force of this argument.

Quite recently, papyri have been discovered which give us information as to the ideas, the beliefs, and the aspirations of a period even earlier than that of the Great Pyramid. The result of the study of these and other records of early Egypt is thus stated by Professor Adolf Erman in *The Historian's History of the World*:

"But when one considers the ancient resident of the valley of the Nile as a human being, with desires, emotions, and aspirations almost precisely like our own; a man struggling to solve the same problems of practical Socialism that we are struggling for today--then, and then only, can the lessons of ancient Egyptian history be brought home to us in their true meaning, and with their true significance. And clearest of all will that significance be, perhaps, if we constantly bear in mind the possibility that the whole sweep of Egyptian history, during the three or four thousand years that separated the Pyramid builders from the contemporaries of Alexander, was a time of national decay--a dark age, if you will--in Egyptian history."

That a great historian, from a study of the ideas and social aspirations of the earliest known civilizations, should have arrived at similar views as to the identity of their mental capacity with our own as I have deduced from their scientific attainments, must be held to be a very strong argument in support of the accuracy of our independent conclusions.

CHAPTER V: SPEECH AND WRITING AS PROOFS OF INTELLIGENCE

There is yet another proof that the faculties of mankind at a very early epoch were fully equal to those of our own time. There is perhaps nothing more difficult in its nature, more utterly beyond the mere lower animal, than the faculty of articulate speech possessed by every race of mankind. We cannot but believe that its acquisition was an extremely slow process, and that it is rendered possible by special cerebral developments giving the necessary mental power for its acquirement.

How long a process this would be, it is impossible to say, but it would certainly have had to reach a high degree of perfection before the equally difficult process of inventing a mode of writing could have been brought to such perfection as to facilitate the further development of the higher faculties through poetry on the one hand and the preservation of facts and discoveries, as well as trains of reasoning, on the other.

Now, I wish to call attention to the very important fact that the origin and development of speech, and later, of writing, were apparently almost simultaneous, and certainly quite independent of each other, in countries not very distant apart. This is shown by the radical diversity of the different groups of languages in Europe, Eastern Asia, and North Africa, and the equal diversity of Egyptian, Assyrian, and Chinese writing. All other written characters are believed to be derived from one or other of these, and it is known that the forms and peculiarities of alphabetic characters have been greatly modified by the various materials employed, such as wood and stone slabs, clay, or wax; papyrus, paper, or parchment; and whether engraved, impressed, or painted, whether written with a reed or quill pen, or with a small brush.

But if intellectual man as a species of mammal had developed by the preservation of variations of survival-value, we should expect to find such an important faculty as speech to have originated in one centre and to have spread rapidly over the world with only slight modifications in isolated communities. The fundamental diversities we find seem to accord better with the conception that when, as a mere animal, his material organism had reached the required degree of perfection, there occurred the spiritual influx which alone enabled him to begin that course of intellectual and moral development, and that marvellous power over the forces of Nature, in which speech and writing, followed by printing, have been such important factors.

In order for man to develop speech, he must have possessed a brain and an intellect far above that of the brutes. As in the more fundamental problem of the origin of life, it is admitted that organization is a product of life--not life of organization; so we must believe that speech was a product of a brain and an intellect sufficient for their development. But such brain and intellect were not necessary for the lower animals, which have reached their highest lines of development in the dog, horse, elephant, and ape without making any definite approach to the acquirement of such higher faculties.

CHAPTER VI: SAVAGES NOT MORALLY INFERIOR TO CIVILIZED RACES

If the facts and arguments set forth in the preceding chapters are correct we should not expect to find any living examples of the unspiritualized man, since the assumption is that the whole race received the influx which started them on their course of purely human development within a strictly limited period, perhaps of a very few generations, or even one generation. The ancestral form--the supposed missing link--would then have become extinct.

If this were not so we should expect to find some isolated groups of speechless man, and of this there is no example; but, on the contrary, the very lowest of existing races are found to possess languages which are often of extreme complexity in grammatical structure and in no way suggestive of the primitive man-animal of which they are supposed to be surviving relics. So long as we got our knowledge respecting them from the low-class Europeans who captured them for slaves or shot them down as wild beasts, we could not possibly acquire any real knowledge of them as human beings. But now that we have more trustworthy accounts of them by intelligent travellers or missionaries, we find ample evidence that when by kindness and sympathy we penetrate to their inner nature, we discover that they possess human qualities of the same kind as our own. A few examples of what unprejudiced witnesses say of them will be very instructive.

Darwin, after attending a meeting between Captain Fitzroy and the chief of a small island near Tahiti to settle a question of compensation for injury to an English ship, says: "I cannot sufficiently express our surprise at the extreme good sense, the reasoning powers, moderation, candor, and prompt resolution which were displayed on all sides."

Captain Cook himself, who saw them in their primitive condition, speaks of the natives of the Friendly Isles as being "liberal, brave, open, and candid, without either suspicion or treachery, cruelty, or revenge"; and a century later Admiral Erskine remarks that "they carry their habits of cleanliness and decency to a higher point than the most civilized nations"; while all the Polynesian races are kind and attentive to the sick and aged, and unlimited hospitality is everywhere practised by them.

Even the Australian aborigines, who are often said to be one of the lowest of human races, are found to possess many good qualities by those who know them best. Mr. Curr, who was for forty years protector of the aborigines in Victoria, says:

"Socially, the black is polite, gay, fond of laughter, and has much *bonhomie* in his composition. . . . The natives are very strict in obeying their laws and customs, even under great temptation. The horror of marrying a woman within the prohibited degrees of relationship, the extreme grief they manifest at the death of children or relatives, and sometimes even for white men, as illustrated by the native boy who was the sole companion of the unfortunate Kennedy when he was murdered, are sufficient to indicate that they possess affections and a sense of right and wrong not very different from our own."

The fact that the physical characteristics of the Australians are substantially those of the Caucasian race in its lowest types has led me to conclude that these interesting people may have been descended from much more civilized remote ancestors, and are thus an example of degradation rather than of survival.* [*See my *Australia and New Zealand*, Chap. V., "The Australian Aborigines," where this view was first set forth. (Stanford, 1893.) For cases of *morality* among savages see my *Natural Selection and Tropical Nature*, p-201.]

Many other illustrations of both intelligence and morality are met with among savage races in all parts of the world; and these, taken as a whole, show a substantial identity of human character, both moral and emotional, with no marked superiority in any race or country. In intellect, where the greatest advance is supposed to have occurred, this may be wholly due to the cumulative effect of successive acquisitions of knowledge handed down from age to age. Euclid and Archimedes were probably the equals of any of our greatest mathematicians of today, while the architecture of Greece, of India, and of Central America is little inferior to mediæval Gothic. But none of these, though so different in style, can be said to prove any real advance in intellectual power from that of the builders of the much more ancient temples and pyramids of Egypt. This latter country, too, in its high material civilization and its remarkable religious system, shows itself the equal of any that has succeeded it.

CHAPTER VII: A SELECTIVE AGENCY NEEDED TO IMPROVE CHARACTER

The general result of the facts and arguments now set forth in the merest outline leads us to conclude that there has been no definite advance of morality from age to age, and that even the lowest races, at each period, possessed the same intellectual and moral nature as the higher. The manifestations of this essentially human nature in habits and conduct were often very diverse, in accordance with diversities of the social and moral environment. This is quite in accordance with the now well-established doctrine that the essential character of man, intellectual, emotional, and moral, is inherent in him from birth; that it is subject to great variation from individual to individual, and that its manifestations in conduct can be modified in a very high degree by the influence of public opinion and systematic teaching. These latter changes, however, are *not* hereditary, and it follows that no definite advance in morals can occur in any race *unless there is some selective or segregative agency at work.*

As there is a great amount of misconception on this subject some explanation may be advisable. Many well-educated and intelligent persons seem to think that whatever characters or faculties are hereditary are also necessarily cumulative. They hear that mental as well as physical characteristics are hereditary; their own observation tells them that there are musical families as well as tall families. They hear that the late Sir Francis Galton wrote a book on *Hereditary Genius*, and perhaps they have read it; but they do not observe that neither he nor any one else has proved that genius of any kind is cumulative, that is, that a man or woman of genius will have, on the average, some one or more children with a greater amount of that special power or faculty than their own. The very contrary of this is really the case. The more a person's talent or mental power is above the average the less chance there is that any of his or her children will have still more of that power than he has. A really great poet, or painter, or musician appears suddenly in a family of mediocre ability or of no ability at all in that special direction. A few examples may be instructive.

Sir William Herschel was the son of a German musician and was himself a musician by profession; but he became an astronomical genius, one of the greatest of his age. His son, Sir John Herschel, was a very clever man, with advantages of education and position. He followed his father as an astronomer, and was a great mathematician, but is never considered to be equal to his father. Darwin's most eminent son was a mathematician, not a naturalist.

The reason of this is that heredity follows the law of

"recession to mediocrity." This is, that all groups of living things vary around an average or mean as regards each of their characters; and those near the average are always numerous, while as we approach the extremes in either direction the numbers become less and less. Families follow the same law. If you take a family for three or four generations, including perhaps some hundreds of persons, some will be short, some tall, but the majority will be near the mean, and the tallest of all will be less likely to have taller descendants than themselves than those nearer the average. But the children of the tallest, though generally shorter than their parents, will still tend to be above the average height.

When a character is so useful to its possessor in the struggle for existence as to be of what is termed "survival value," then those that vary most above the average will be preserved or selected generation after generation as long as the increase is useful.

It is because the higher intellectual or moral powers are so rarely of life-preserving value, and are not infrequently the reverse, that they are not *cumulative*, though they are *hereditary*.

With this explanation we will now proceed to examine somewhat closely our moral position as a nation; what is the nature of our social environment; how it came to be what it is, and what lessons we may learn from it.

CHAPTER VIII: ENVIRONMENT DURING THE NINETEENTH CENTURY

During the eighteenth century our material civilization, which had long been almost stationary, began to advance with the growth of the physical sciences, but at first with extreme slowness. The earliest steps were made by the application of machinery to some of the domestic arts. Some refinements were made in the manners and customs of our daily life; but there were few, if any, indications of permanent or widespread change, either for better or worse, in our intellectual or moral nature.

The nineteenth century, however, saw the initiation of a great change in the economic environment due to the rapid invention of labor-saving machinery, which, with the equally rapid application of steam power, led to an increase of wealth production such as had never been known on the earth before. During the same period new modes of locomotion were brought into daily use, the facilities for inter-communication were increased a hundredfold, scientific discoveries opened up to us new and unthought-of mysteries of the universe, and the whole earth was ransacked for its treasures, both vegetable and mineral, to an extent that surpassed all that had been accomplished since the dawn of civilization.

But this rapid growth of wealth, and increase of our power over Nature, put too great a strain upon our crude civilization and our superficial Christianity, and it was accompanied by various forms of social immorality, almost as amazing and unprecedented. Some of these may be here briefly referred to.

Our vast textile factory system may be said to have commenced with the nineteenth century, and the profits were at first so large and so dependent on the supply of labor that the mill-owners hired children from the workhouses of the great cities by hundreds and even thousands. These children, from the age of five or six upwards, were taken as apprentices for seven years, and they really became the slaves of the manufacturers, whose managers made them work from 6 a.m. to 7 p.m., or sometimes longer; and, in order to keep them awake in the close atmosphere of the factories it was found necessary to whip them at frequent intervals. It was not till 1819 that the age of children employed in factories was raised to nine years, while in 1825 the working hours were *limited* to seventy-two a week!

From that time onward, during the whole of the nineteenth century, there was a continued succession of "Factory Acts," each aiming at abolishing or ameliorating the worst results of child labor--its inhumanity, its cruelty, and its immorality. These legislative efforts were always opposed by the employers, who usually succeeded in so mutilating them in Committee of the House of Commons as to render them almost useless. Mrs. E. B. Browning's noble verses, *The Cry of the Children*, show that after nearly fifty years of struggle the condition of the child-workers was still, in a high degree, cruel, degrading, and therefore immoral, while that of the half-timers who succeeded them was almost as injurious.

As the century wore on, other evils of a similar nature were gradually brought to light. Children and women were found to be working underground in coal mines, under equally vile conditions as regards health and morality; and an enormous loss of life was caused by inadequate ventilation, insecure roof-propping, imperfect winding machinery, and other causes, all due to want of proper precautions by the owners of the mines. As a matter of simple justice, such owners should be held responsible to the injured person not only to the full extent of his wages and for medical attendance, but should also pay a liberal compensation for the pain suffered, and for the extra labor, expense, and anxiety to his family. But all such things are ignored in the case of poor workers, so that even the money compensation is reduced to the smallest amount possible.

It is one of the great defects of our law that deaths due to preventable causes *in any profit-making business* are not criminal offences. Till they are made so, it will be impossible to save the hundreds, or even thousands, of lives now lost owing to neglect of proper precautions in all kinds of dangerous or unhealthy trades. However costly such precautions may be, expense should not be considered when human life is risked; and the present state of the law is therefore immoral.

Notwithstanding Acts of Parliament and numerous Inspectors (whose salaries should be paid by the mine owners), explosions and other accidents underground continue to increase, the year 1910 being a record year, with its 1,775 deaths; and even the number in proportion to the workers employed is the highest for the last twenty years.

Yet no one is punished, or even held responsible for these deaths. Surely, this shows a deplorable absence of moral feeling, both in the general public and in Parliament. The responsibility of Parliament is really criminal, since it always allows its legislation to be made ineffective by the fear of diminishing the employers' profits, thus deliberately placing money-making above human life and human well-being.

In the case of mines and quarries, Parliament is especially responsible, because the possession of the mineral wealth of our country by private individuals is itself a gross usurpation of public rights, and should have been long ago declared illegal. Whatever arguments--and they are very strong--show us that the land itself should not be private property, are ten times stronger in the case of the minerals within its bowels.

The value of land increases with its proper use, but in the case of minerals, the value is absolutely destroyed. Surely, it is a crime against posterity to allow the strictly limited mineral wealth of our country to be made private property, and very largely sold to foreigners, solely to increase the wealth of individuals and to the absolute impoverishment of ourselves and our children.* [*I pointed this out forty years ago in an article entitled *Coal a National Trust*, which I republished twelve years ago in my *Studies, Scientific and Social* (Vol. II., Chap. VIII.).]

I will here add one other argument which goes to the root of the matter by showing that the alleged owners of minerals have not even a legal title to them. It is, I believe, a maxim of law that public rights cannot be lost by disuse. Landed estates were, in our country, created by the Norman Conqueror to be held subject to the performance of feudal duties. Deep-seated minerals were then not known to exist, and were not (I believe) specifically included in the original grants. Except, therefore, where they have since been made private property by *Act of Parliament*, they still remain public property. I submit, therefore, that they may be both legally and equitably resumed by the Government as public property, and worked for the good of the public and of posterity. Compensation to the supposed present owners would be a matter of favor, *not of right*.

CHAPTER IX: INSANITARY DWELLINGS AND LIFE-DESTROYING TRADES

The enormous difference between town and country dwellers as regards duration of life and the prevalence of zymotic diseases has been known statistically since the era of registration, and a body of Health Officers has been set up to report upon the worst cases. The local authorities have power to compel the owners of unhealthy dwellings to put them into a sanitary condition, or even order them to be entirely rebuilt. But as many of the members of corporations and other local boards are often themselves owners of such property, or have intimate friends who are so, very little has been done to remedy the evil. Again and again, in all parts of the country, the Health Officers have duly reported, but their reports have been ignored. In some cases, where the Health Officer has been too persistent, he has been asked to resign or has been discharged. A few general facts may be here given.

By the last complete Census returns (1901), there are in England and Wales 7,036,868 tenements, and of these 3,286,526, or nearly half, have from one to four rooms only. In London, out of a total of 1,019,646 tenements, 672,030, or considerably more than half, have from one to four rooms; while there are about 150,000 tenements of only *one room*, in which are living 313,298 persons, or about two and a quarter persons in each room on the average. There are, however, about 20,000 persons living *five in a room*, and 20,000 more who have *six, seven, or eight in a room*. As most of these one-roomed tenements are either the cellars or attics of houses in the most crowded parts of large towns, where there is impure air, little light, and scanty water supply, the condition of those who dwell in them may be imagined--or rather *cannot* be imagined, except by those who have explored them.

Equally inhuman, immoral, and even criminal, is the neglect of all adequate measures to check the loss of infant life through the overwork, poverty, or starvation of the mother, together with overcrowded and insanitary dwellings. In the mad race for wealth by capitalists and employers most of our towns and cities have been allowed to develop into veritable death-traps for the poor. This has been known for the greater part of a century, yet nothing really effective has been done, notwithstanding abundant health legislation--again made useless by the dread of diminishing the excessive profits of manufacturers and slum-owners. One of the Labor newspapers calls our attention to the following facts for 1911 as to Infant mortality per 1,000 born:

Deptford, East Ward (poor) . . . . 197 per 1,000

Deptford, West Ward (rich) . . . . 68 per 1,000
Bournville Garden Village . . . . 65 per 1,000
St. Mary's Ward, Birmingham . . . . 331 per 1,000

Such facts exist all over the kingdom. They have been talked about and deplored for the last half-century at least. Who has murdered the 100,000 children who die annually before they are one year old? Who has robbed the millions that just survive of all that makes childhood happy--pure food, fresh air, play, rest, sleep, and proper nurture and teaching? Again we must answer, our Parliament, which occupies itself with anything rather than the immediate saving of human life and abolishing widespread human misery, the whole of which is remediable. And all for fear of offending the rich and powerful by some diminution of their ever-increasing accumulations of wealth. No thinking man or woman can believe that this state of things is absolutely irremediable; and the persistent acquiescence in it while loudly boasting of our civilization, of our science, of our national prosperity, and of our Christianity, is the proof of a hypocritical lack of national morality that has never been surpassed in any former age.

A new set of evils has grown up in the various so-called "unhealthy trades"--the lead glaze in the china manufacture, the steel dust in cutlery work, and the endless variety of poisonous liquids and vapors in the numerous chemical works or processes, by which so many fortunes have been made. These, together, are the cause of a large direct loss of life, and a much larger amount of permanent injury, together with a terrible reduction in the duration of life of all the workers in such trades. Yet in one case only--that of phosphorus matches--has any such injurious process of manufacture been put an end to. Wealth has been deliberately preferred to human life and happiness.* [*An account of some deadly trades is given in Mr. R. H. Sherard's book, *The White Slaves of England.*]

One of the most deadly of trades seems to have remained unnoticed till it has been brought to light by the new Labor paper, *The Daily Citizen*, in a series of articles by Mr. Keighley Snowden entitled *The Broken Women*. Never was a title better deserved, since large numbers of girls and young women are employed at Lye and Cradley Heath, in what is commonly named the "Hollow-ware" works. This is the tinning, or galvanizing, as it is usually termed, of buckets and other domestic utensils, in which lead is used; and it produces one of the most virulent forms of lead-poisoning. The symptoms are, among other more painful ones, the loss of hair and the loosening and ultimate loss of teeth, culminating either in chronic illness or death, sometimes in a few months or years. Five years ago there was a Home Office inquiry, which, after full examination, reported that the process used was dangerous to life, that no precautions could render it harmless, and that it should be *totally discontinued*.

An order was then issued by the Home Office that after a time-limit (two years) the process should be no longer used; but that order has not been obeyed (except by a few employers) to this day. The deadly nature of this work was accompanied by miserably low wages, as shown by the fact that the women workers have at length struck to obtain a minimum of 10 *s.* a week! Helped by some humane friends, they have at length succeeded in obtaining this miserable wage, and for the present are in a state of comparative happiness! How long it will be before the Government abolishes this deadly process we cannot tell. The following is a brief statement of what these poor women have to suffer, extracted from *The Daily Citizen* of November 20, 1912:

"They had, without power to resist them, suffered repeated and ruthless reductions of wages. They had seen their industry brought down by reckless competition, and the manufacture of shoddy goods, to the point at which men could no longer earn enough to support their families. They had seen their wives and daughters and boys forced by want at home into workshops, where, as official inquiry has shown, health was sucked out of their bodies as though they had been the victims of vampires. They had seen the introduction and growth of the sub-contracting 'stint' system, under which boyhood and girlhood and motherhood were driven as though they had been slaves under the lash, and their earnings cut down to a penny an hour. Meanwhile, they lived in the hovels and holes of a place which can only be fitly described as one of the dirtiest ashpits of a civilization reckless of dirt where profit is a question."

Those who want to know what horrors can exist today in England should read Mr. Snowden's series of articles on the subject. They are restrained in language, and state the bare facts from careful personal observation. That such things should still exist in a country claiming to be civilized would be incredible, were there not so many others of a like nature and almost as bad.

In an almost exhaustive volume on *Diseases of Occupation* by Sir Thomas Oliver, M.D. (1908), there is only a short reference to the hollow-ware trade of the "black country" near Birmingham. But the tin-plate industry of South Wales is more fully described, with the same pitiable condition of the women workers and the same terrible results to health and life. Yet nothing whatever seems to be done by the manufacturers; and though two Home Office Inspectors have fully reported on its horrors from 1888 onwards, no notice appears to have been taken of them, nor has there been any Government interference with conditions of labor which are a disgrace to civilization.

CHAPTER X: ADULTERATION, BRIBERY, AND GAMBLING

After the terrible national crime of deadly employments it is almost an anti-climax to enumerate the vast mass of dishonesty and falsehood that pervades our commercial system in every department. Almost every fabric, whether of cotton, linen, wool, or silk, is so widely and ingeniously adulterated by the inter-mixture of cheaper materials that the pure article as supplied to our grandparents is hardly to be obtained. Of this one example only must serve. Calicoes have been successively dressed with such substances as paste and tallow; then with the still cheaper china clay and size; and in some cases from 50 to 90 per cent. of these latter materials have been sold as calico for exportation to countries inhabited by what we term savages. These people only found out the deception when the need for washing or exposure to tropical rains reduced the material to a flimsy and worthless rag, as I have myself witnessed in some parts of the Malay Archipelago.* [*These facts are given in the Ninth Edition of the "Encyclopædia Britannica." In recent editions the article *Adulteration* is limited to food and drugs. In "Chambers' Encyclopædia," cotton, linen, and woollens are included among adulterated fabrics.]

Even worse is the adulteration of almost every kind of prepared food--including the showy sweetmeats which tempt our children--with various chemicals, which are often injurious to health, and sometimes fatal; while even the drugs we take in the endeavor to cure our various ailments are frequently so treated as to be useless or even hurtful. Along with this form of dishonesty is what may be termed simple cheating in the description of goods sold, especially as to quantity. Threads and fabrics are generally shorter or narrower than stated, giving a larger profit when sold in enormous quantities in our great retail shops.

Then, again, there is a widespread system of bribery of servants or other employees in order to obtain more customers or to secure contracts; and though these are all criminal offences, and a great host of inspectors and official analysts are employed to discover and convict the offenders, yet so few people are willing to take the trouble and lose the time and money involved in putting the law into motion, that a very large percentage of these offences go undiscovered and unpunished.

Yet another and more serious form of plunder of the public is carried on by means of Joint Stock Companies, of which there are now more than 50,000 in England and Wales. In the year 1911 the number of new companies was 5,959, while 4,353 ceased to exist, giving an increase of 1,606 in the year. The Limited Liability Act was passed in

1855, in order that the public might invest their savings in companies, and thus share in the profits of our industry and commerce. It was supposed to be quite proper that anyone should benefit by the enterprise and industry of others; but to do so is essentially immoral and has resulted in a vast system of swindling and terrible losses to the innocent investors. The promoters, directors, secretaries, and bankers of these companies always gain; those that take up the shares often lose; and the amount of misery and absolute ruin of those who fondly hoped to add to their scanty incomes, and have been deluded by the names of well-known public men among the directors, is incalculable.

Our Stock Exchanges, too, are used largely for pure gambling, which, owing to its vast extent and being carried on under business forms, is perhaps more ruinous than any other. But this form of gambling goes on unchecked, and is generally accepted as quite honest business. Yet ordinary betting on races and other forms of direct gambling are hypocritically condemned as immoral and criminal.

The vast fabric of our foreign trade in food, or the raw materials of our manufactures, is also used to support perhaps the greatest system of gambling the world has ever seen. The fluctuating prices of corn or cotton, of coal or mineral oil, of iron and other metals, in the great markets of the world, are used in two ways by a large community of gamblers, who not only do not require the goods they buy, but who never see nor possess them. The ordinary speculator who buys when prices are low, to sell again at a profit, without himself being able to influence the rise or fall of price, is a pure gambler who thinks he can foresee the changes of the market price in the immediate future. But the great capitalists who, either singly or by means of what are called rings or combines, purchase such vast quantities of the special product as to create a scarcity in the market, leading to a large rise of price, are ingenious robbers rather than gamblers, because, by clever dealings with such a monopoly, often aided by false rumors widely circulated in newspapers owned or bribed by them, they are able to make enormous profits at the expense of those who are obliged to purchase for actual business purposes or for daily use. This is one of the methods by which the great millionaires and multi-millionaires of the world accumulate their wealth, every penny of which is at the cost of the consuming public.

This is certainly as immoral as any of the petty forms of swindling with marked cards, loaded dice, or the wilful losing of a race; yet the possessors of such wealth are usually held to be clever business men, whose morality is not questioned.

All these inconsistencies as regards the moral status of various kinds of gambling or dishonest speculation arise from our inveterate habit of dealing with limited cases, each judged on its supposed merits as to consequences, instead of looking to fundamental principles. Why is gambling immoral? Not because it is a game of chance entered into for mere amusement, even when played for small money stakes which are of no importance to any of the players. The fundamental wrong arises whenever it is used for obtaining wealth or any part of the player's income; and the reason is, that whatever one wins, some one else loses; while its evil nature, socially, depends upon the fact that whoever acquires wealth by such means contributes nothing useful to the social organism of which he forms a part. If it were taught to every child, and in every school and college, that it is morally wrong for any one to live upon the combined labor of his fellow-men without contributing an approximately equal amount of useful labor, whether physical or mental, in return, all kinds of gambling, as well as many other kinds of useless occupation, would be seen to be of the same nature as direct dishonesty or fraud, and, therefore, would soon come to be considered disgraceful as well as immoral.

We see, then, that the whole commercial fabric of our country--our immense mills and factories, our vast exports and imports, our home trade, wholesale and retail, and innumerable transactions in our stock exchanges--is permeated with various forms of dishonesty, gambling, and direct robbery of individuals or of the public. No class is wholly free from it, and it increases in volume from decade to decade, just as our boasted commerce and accumulated wealth increases.

I have here called attention to these various forms of immoral practices because they are so often ignored. Yet they are all officially admitted by the enormous mass of the various Royal Commissions, Parliamentary and other reports, as well as by the hundreds of "Acts" by which successive Parliaments have endeavored to deal with them, but which have, one and all, proved to be either wholly or partially ineffective. The reason of this failure is that in every case symptoms and isolated results only have been considered, while the underlying causes of the whole vast mass of social corruption have never been sought for, or, if known, have never influenced legislation.

CHAPTER XI: OUR ADMINISTRATION OF "JUSTICE" IS IMMORAL

When we read about the Turkish or other Eastern law courts, in which direct bribery of every official up to the judge himself is a regular feature, we are horrified, and are apt to proclaim the fact that our judges never take bribes. But, practically, it comes to very nearly the same thing in England. No single step can be made for the purpose of getting justice without paying fees; while the whole process of bringing or defending an action-at-law is so absurdly complex as to be almost incredible. Jeremy Bentham satirized this by supposing a father of a large family to adopt the same method of settling a dispute between two of his sons. He would not hear either of them himself, but each must tell his story to a stranger (a solicitor), who wrote it down and then instructed another stranger (a barrister) to explain it to the father (as judge) and twelve neighbors (the jury). Then the stranger (barrister) on each side asked questions of all the family who knew anything about it; and the barristers, who had only third-hand knowledge of the facts, tried to make each witness contradict himself, or to acknowledge having done something as bad another time; till the jury became quite puzzled, and often decided as the cleverest of the barristers told them.

That is really the system of law courts to this day; and it is grossly unfair, because the party who can pay the highest fees for the services of the most experienced counsel is most likely, through the lawyer's skill and eloquence, to secure a verdict in his favor. Yet there is no effective protest against this unjust and absurd system, which absolutely denies all redress of wrongs to the poor man when oppressed by a rich one. One would think it self-evident that justice ceases to be justice when it has to be paid for. But the system is so time-hallowed, the profession of a barrister so honored, and its rewards so great, that it will never be abolished till there comes about in our social system that fundamental change which will cut at the very root-cause of almost all our existing lawsuits, immorality, and crime.

In our criminal as well as our civil law and procedure there is equal injustice. When the poor man is accused of the slightest offence and brought before a magistrate by the police he is, even though perfectly honest and respectable, treated from the very first as if he were guilty, often refused communication with his friends; and, when the accusation is serious, he is remanded to prison again and again till evidence has been hunted up, or even manufactured, against him. Experience shows that the latter is often done and a quite innocent man not infrequently punished. The dictum of the law, that an Englishman should be held to be innocent till he is proved to be guilty, is absolutely reversed, and he is treated as if he were guilty till, against overwhelming odds, he is able to prove himself innocent. There is no possible excuse for this now, and at the very least every man who has a home or a permanent employment should be at once discharged on his own recognizances.

Equally unjust and barbarous is the system of money-fines, often for merely nominal offences, with the alternative of imprisonment. To the well-off, or to the habitual criminal, the fine is a trifle; but to the poor man charged with being drunk, with begging, or with sleeping under a haystack, or any such act which is no real offence, the common punishment of 10 *s.* or a week's imprisonment, leaving perhaps wife and children to starve or be sent to the workhouse, is really far more immoral than the alleged offence.

Again, our Poor Law itself, as usually administered, is utterly immoral. This is what a competent authority--Mr. Sidney Webb--says of it:

"Underneath the feet of the whole wage-earning class is the abyss of the Poor Law. I see before me a respectable family applying for relief. What do we do to them? We, the Government of England, break up the family. We strip each individual of what makes life worth living. When the man enters the workhouse he is stripped of his citizenship--branded as too infamous to vote for a member of Parliament. Once in the workhouse, we put him to toil or to loiter under conditions that are so demoralizing that we turn him into a wastrel. And we strip the wife of her children. We send her to the wash-tub or the sewing-room, where she associates with prostitutes and imbeciles. The little children, if they are under five, are taken to the workhouse nursery, where they also are tended by prostitutes and imbeciles. There they remain, day after day, without ever going down the workhouse steps until they are old enough to go to the Poor Law school, or until they are taken down in their coffins, owing to the terrible mortality among the workhouse babies."

Of course, all workhouses are not so bad as this, but many are, and have been during the three-quarters of a century of their existence. Can we, therefore, wonder that week by week some poor and honest parents commit suicide rather than see their children starve, or be separated from them in the workhouse! The people we thus drive to death are many of them as good as we ourselves are; yet the "Guardians of the Poor"--well-to-do gentlemen and ladies--go on administering it week after week and year after year without protest or apparent compunction. Such is the deadening effect of long-continued custom.

CHAPTER XII: INDICATIONS OF
INCREASING MORAL DEGRADATION

There are in the Reports of the Registrar-General a few statistics of special importance because they clearly point to certain kinds of moral degradation which have been increasing for the last half-century, thus coinciding with our exceptionally rapid increase in wealth; and also, as I have shown in preceding chapters, with various forms of national, economic, and social deterioration.

The first of these is the continuous increase in deaths from alcoholism, in proportion to population, since the year 1861. Most persons will be amazed to find that this is the case, because the drinking habit has certainly diminished; but when the habit becomes so powerful and lasts so long as to be the direct cause of death, we are able to see the dimensions of the most exaggerated form of the drink evil. The following figures are taken from the successive Reports referred to:--

Average of Years	Deaths from Alcoholism per Million living
1861-1865	41.6
1866-1870	35.4
1871-1875	37.6

1876-1880	42.4
1881-1885	48.2
1886-1890	56.0
1891-1895	67.8
1896-1900	85.8
1901-1905	78.4
1906-1910	54.6

There are some irregularities, the ratio being nearly equal for the first twenty years, after which there is such a continuous large increase that from 1876-80 to 1896-1900 the mortality is doubled, but for the last ten years there has been a decrease, which in the last five years is very marked.

But a still worse and more disquieting feature is the recent large increase of mortality from alcoholism in women. Figures for the separate sexes were not given till 1876, and the following table shows the comparison up to 1910:--

Average of Years	*Deaths from Alcoholism per Million*
1876-1880	60.1 (Men) 24.0 (Women)
1881-1885	66.6 (Men) 31.0 (Women)
1886-1890	73.6 (Men) 39.2 (Women)
1891-1895	86.6 (Men) 50.2 (Women)
1896-1900	106.2 (Men) 66.6 (Women)
1901-1905	95.0 (Men) 63.0 (Women)
1906-1910	66.6 (Men) 43.6 (Women)

These figures, however deplorable and startling in themselves, are as nothing in comparison with what they imply. Death from drink, more than in the case of any other disease, is the ultimate and rarely attained result of the vice of habitual intoxication. Men and women may greatly injure their health, ruin their families, and be disgraceful drunkards, and yet not die of it, or make any near approach to doing so. What is the proportion of those who are morally and physically injured by drink to those who kill themselves by it, is, I suppose, unknown, but I imagine that one in a thousand is, probably, too high an estimate, and that one death among ten thousand moderate drinkers who also occasionally or frequently become intoxicated, would be nearer the mark. This would imply an increase in the consumption of alcoholic drinks, instead of which there has been an actual diminution. The fact probably is that a very large number of moderate drinkers have ceased to consume alcohol in any form, and this would account for a much larger reduction in the total than has actually occurred.

On the other hand, owing to the increase of those who are only casually employed in our great cities, and whose one luxury is the excitement of drink, a larger quantity of cheap and injuriously adulterated spirits and other liquors is consumed, which, combined with a deficiency of wholesome food, leads more frequently to a fatal result.

Increase of Suicide

The increase has been long known and generally admitted. It is supposed to be largely due to the ever-increasing struggle for subsistence in our great cities, the consequent increase of unemployment, and the dread of the workhouse as the only alternative to starvation. The following are the figures for the last forty-five years for which official data have been published:--

Average of Years	*Deaths by Suicide per Million living*
1866-1870	66.4
1871-1875	66.0
1876-1880	73.6
1881-1885	73.8
1886-1890	79.4
1891-1895	88.6
1896-1900	89.2
1901-1905	100.6
1906-1910	102.2

Such a table as this, occurring in a country which boasts of its enormous wealth, of its ever-increasing commercial prosperity, of its marvellous advance in science and the arts, and command of natural forces, should, surely, give us pause and force upon us the conviction that there is something radically wrong in a social system which brings about such terrible evils.

And this should be the more certainly seen to be the case because the same increase is taking place in all those countries which approach us in their wealth and their commercial prosperity.

There is a group of diseases which are fatal to infants soon after birth. They have been steadily increasing during the last half-century, and call for special notice here, as they seem to indicate physical degeneration as well as personal immorality of a dangerous and perhaps even a criminal nature.

Five-year Average Proportion
of Deaths to 1,000 Births

Premature Births	Congenital	Defects
1861-1865	11.19	1.76
1866-1870	11.50	1.84
1871-1875	12.60	1.85
1876-1880	13.38	2.39
1881-1885	14.18	3.23
1886-1890	16.1	4.2
1891-1895	18.4	4.7
1896-1900	19.6	4.9
1901-1905	20.2	5.9
1906-1909	20.0	6.6

The large increase during the last forty-five years of very early infantile deaths, involving abnormalities of mother or child, seems very significant. The first may be connected with the increasing dislike of child-bearing, and unsuccessful attempts to avoid it. The second indicates some injurious condition of life of the mother, such as working at unhealthy or even deadly trades, which has certainly been largely increasing during the same period. Such work for young married women should be impossible in a civilized community.

On the vast subject of prostitution, of which the present movement for the suppression of what is called "The White Slave Traffic" is but one of the aspects, I do not propose to dwell, because I can find no statistics to show whether it has increased or decreased during the last century. But as the conditions have all been favorable for it, I have little doubt that it has increased in proportion to population. Such conditions are, the enormous growth of great cities; an increasing number of unmarried and wealthy young men; with an enormous number of girls and young women whose wages are insufficient to provide them with the rational enjoyments of life.

The proceedings of the divorce courts show other aspects of the result of wealth and leisure; while a friend who had been a good deal in London society assured me that both in country houses and in London various kinds of orgies were occasionally to be met with which could hardly have been surpassed in the Rome of the most dissolute emperors.

Of war, too, I need say nothing. It has always been more or less chronic since the rise of the Roman Empire, but there is now undoubtedly a disinclination for war among all civilized peoples. Yet the vast burden of armaments, taken together with the most pious declarations in favor of peace, must be held to show an almost total absence of morality as a guiding principle among the governing classes. In this respect, the increasing power of labor-parties all over the world seems to afford the only hope of a real moral advance.

PART II.--THEORETICAL

CHAPTER XIII: NATURAL SELECTION AMONG ANIMALS

While writing the present volume I was led to refer to it during some of the numerous interviews on the occasion of my recent birthday. This led to some misrepresentation of my views, and showed me how few popular press-writers have any real knowledge of the nature and extent of "natural selection," more especially as it affects the human race. There is also the same ignorance as regards "heredity"; and this latter has become almost a word to conjure with, and is thought by most writers to explain many things to which it is quite inapplicable, and as the present work is a very condensed argument founded to a considerable extent upon these great natural laws, I propose devoting two chapters to explaining and demonstrating the effect of natural selection in the case of the lower animals and of man respectively.

That such an explanation is necessary may be seen from the following extract from one of our most influential and well-written daily papers, the *Pall Mall Gazette*. After referring to the view of the utter rottenness of our present civilization, it quotes me as saying: "And the average of mankind will remain the same until natural selection steps in to save it." (What I actually said to the interviewer was "until some form of selection improves it.") The writer then goes on:

"These words must have struck the interviewer like the crack of doom. For, stated popularly, the theory of natural selection is the doctrine of 'Devil take the hindmost.' If natural selection had fair play there would be no Children's Care Committees; there would be no Poor Law, no Hospitals; there would be no Old Age Pensions. All the humanitarian effort to care for the weak and to help them along the path of life, every effort to bind up the broken-hearted, every combination of labor to secure equality among the members of a trade, stand condemned as futile or worse by the doctrine which Dr. Russel Wallace thinks can alone raise the average of man. His own remedies for the ills of society--the levelling up which he believes to be impossible without levelling down, the disinheriting of the unborn heir, the 'striking' which he applauds, the universal education which he favors--all these are directly antagonistic to the workings of natural selection."

Now, as I am credited by all my scientific friends with having discovered the theory of natural selection more than fifty years ago, and as the whole reading public have had this hammered into them with needless repetition during the whole of that period, it is rather amusing to be told now that I do not know what natural selection is, nor what it implies. It is also a striking proof that the whole subject is now held to be so old and commonplace as not to be worth studying by a popular teacher before writing about it so strongly and dogmatically. If he had done so he would not deliberately assert that I hold opinions in regard to the matter which in several of my books I have shown the fallacy of.

I propose, therefore, to give here a short account of the essential features of the theory of natural selection; how it has operated in bringing about the evolution of the almost infinitely varied forms of plants and of the lower animals; and also to explain as clearly as I can why, and to what extent, it has acted differently in the case of man.

Lamarckism and Darwinism--How they Differ

The first great naturalist who put forward a detailed explanation of how he supposed the varied forms of animal life to have been produced was Lamarck, a contemporary of Buffon and Goethe, both of whom believed in evolution, but offered no explanation of how it could have been brought about. Lamarck, however, suggested that the various organs of animals were modified by voluntary effort producing increased development, as when an antelope escapes from a lion by its swiftness, which swiftness is increased by the straining of its limbs in flight; while the long neck and fore-limbs of the giraffe were explained by the continual stretching of these parts of the body to obtain foliage for food during severe droughts. In addition to this other causes are at work, as described in the following passage, translated or paraphrased by Sir Charles Lyell in his *Principles of Geology*:

"Every considerable alteration in the local conditions under which each race of animals exists causes a change in their wants, and these new wants excite them to new actions and habits. These actions require the more frequent employment of some parts before but slightly exercised, and then greater development follows as a consequence of their more frequent use. Other organs, no longer in use, are impoverished and diminished in size; nay, are sometimes entirely annihilated, while in their place new parts are insensibly produced for the discharge of new functions."

Again, he says:

"Thus otters, beavers, water-fowl, turtles, and frogs were not made web-footed in order that they might swim; but their wants having attracted them to the water in search of prey, they stretched out the toes of their feet to strike the water and move rapidly along its surface. By the repeated stretching of their toes the skin which united them at the base acquired a habit of extension, until, in the course of time, the broad membranes which now connect their extremities were formed."

In the case of plants, where no voluntary movements occur, the cause of modification was said to be due almost exclusively to the change of local conditions, as the various kinds of plants became dispersed over the earth's surface. The influence of soil, of temperature, of light and shade, are supposed to produce definite changes which are gradually increased; just as plants long cultivated in our gardens have become so changed that the wild progenitors cannot now be recognized.

Sir Charles Lyell, who made a careful study of Lamarck's great work, notes especially that the whole of the argument is vague and general, and that no cases are given in which is shown how the alleged causes can be supposed to have acted so as to bring about the innumerable changes that must have occurred. What is more important, however, is the failure to explain how the numerous minute adaptations of each species to its environment could have arisen by the direct action of that environment--in plants, the infinitely varied forms of leaves, flowers, and fruits; in animals, the forms and sizes of the teeth of mammalia and of the beaks, wings, and feet of birds to the food they obtain; while the enormous range of color and marking in most groups of animals are such as no amount of desire or exertion on the one hand, or direct action of external causes on the other, could possibly have brought about. It is not, therefore, surprising that, although a vast amount of evidence was adduced to show that changes had taken place leading to the evolution of species from pre-existing species, yet causes adequate to bring about the changes, and especially those necessary to produce the marvellous adaptations continually being discovered, had not been shown to exist.

It is necessary to point this out, because the difference between the almost universal rejection of Lamarck's attempted solution of the problem of evolution, and the almost immediate and universal acceptance of that adduced by Darwin, is otherwise unexplained. The belief in the doctrine of evolution as the only rational explanation of the gradual development of the innumerable forms of living things became more and more general. The great body of arguments in its favor were admirably set forth by Robert Chambers in his *Vestiges of Creation*, published anonymously in 1844; while Herbert Spencer's masterly exposition of the argument for universal evolution convinced a large number of naturalists and men of science. But still the nature of the laws and forces by which the evolution of the organic world, in all its variety and beauty, could have been brought about remained not only unknown but unimagined, so that even so great a thinker as Sir John Herschel termed it "the mystery of mysteries." I will now state as briefly as possible the essential features of Darwin's solution of the mystery in his epoch-making work, *The Origin of Species*.

Natural Selection as the Essential Factor in the Origin of Species

There are two great, universal, and very conspicuous characteristics of the whole organic world which, because they are so very common, were almost ignored before Darwin showed their importance. These are (1) the great *variability* in all common and widespread species, and (2) their enormous powers of *increase*.

The facts of variability are recorded in every book on Darwinism or on organic evolution, and it is only necessary here to appeal to the reader's own observation or to state a few illustrative facts. Everybody sees that among a hundred or a thousand people he knows or frequently meets no two are alike. This is *variability*. He also knows that the amount of the differences between them is often very large, and always, if you have any two of them side by side, easily perceptible and capable of being described. He also knows that they differ in every part and organ that can be seen: the height, the bulk of body, the shape of the hands, feet, head, ears, nose, and mouth; the proportions of the legs, arms, and body to each other; the abundance and character of the hair--coarse or fine, straight or curly, and of all colors between flaxen and intense black. To declare that variability among men and women, even of the same race and in the same country, is a rare phenomenon, and that in amount it is infinitesimal, would be a ludicrous mis- statement of the facts or a wilful perversion of the truth. But, as regards animals or plants in a state of nature, this misstatement has been made and has been used as an argument against the Darwinian theory. It is, however, now well known, as a matter of direct observation and measurement, that when a few scores or hundreds of individuals are compared, even in the same district and at the same season, they differ in their proportions to about the same amount, and to some extent in every visible part or organ, as do human beings.

This, however, was not well known when Darwin collected the materials for his various works, and he even sometimes makes the proviso, "if they vary, for without variation selection can do nothing"; and this has been taken as an admission that variation is a rare instead of being a universal phenomenon. He also often spoke of the accumulation of *small* or *minute* variations, and this has led to the statement that variations are *infinitesimal* in amount, and therefore could, at first, be of no use to the possessor in the struggle for existence.

Rapid Increase of All Organisms

This is another fact of Nature which requires to be kept in mind in all discussions of the action of natural selection, yet it is often altogether ignored by critics of the theory. As an illustrative fact, a not uncommon European weed of the Cruciferæ family has been found to produce about 700,000 seeds on a single plant, whence it can be calculated that if every seed had room to grow for three successive years their produce would cover a space of about 2,000 times as large as the whole land surface of the globe. Some of the minute aquatic forms of life which increase by division in a few hours would, if they all had the means of living, in the same period occupy a space equal to that of the entire solar system. Even the largest and slowest breeding of all known mammals, *i.e.,* the elephant, would, if allowed space to live and breed freely for 750 years, result in no less than nineteen million animals.

By far the larger part of the criticisms of Darwinism by popular writers are due to their continually forgetting these two great natural facts: enormous *variability* about a *mean value* of every part and organ, and such ever-present powers of multiplication that, even in the case of vertebrate animals, of those born every year only a small proportion--one-tenth to one-hundredth or thereabouts--live over the second year. If they all lived their numbers would go on continually increasing, which we know is not the case. Hence arises what has been termed "the struggle for existence," resulting in "the survival of the fittest."

This "struggle for life" is either against the forces of inorganic or those of organic nature. Among the former are storms, floods, intense cold, long-continued droughts, or violent blizzards, all of which take toll of the weaker or less wary individuals of each species--those that are less adapted to survive such conditions. In judging how this would act, we must always remember the enormous scale on which Nature works, and that, although now and then a few of the weaker individuals may live and a few of the stronger be killed, yet when we deal with hundreds of millions, of which eighty or ninety millions inevitably die every year, while about ten or twenty millions only survive, it is impossible to believe that those which survive, not one year only, but year after year throughout the whole existence of each species, are not on the average better adapted to the complex conditions of their environment than those which succumb to it. It is a mere truism that the *fittest survive.*

Exactly the same thing occurs in the case of the organic environment, to which each species must also be well adapted in order to live. The two great essentials for animal existence are, to obtain abundant food through successive years, and to be able to escape from their various enemies. When food is scarce the strongest, or those who can feed quickest and digest more rapidly, or those that can detect food at greater distances or reach it more quickly, will have the advantage. Enemies are escaped by strength, by swiftness, by acute vision, by wariness, or by colors which conceal the various species in their natural surroundings, and those which possess these or any other advantages will in the long run survive. The weaker, the less well-defended, and the smaller species often have special protection, such as nocturnal habits, making burrows in the earth, possessing poisonous stings or fangs, being covered with protective armor, while great numbers are colored or marked so as exactly to correspond with their surroundings, and are thus concealed from their chief enemies.

Natural Selection, or Survival of the Fittest

It may be here noted that the term "natural selection," which has often been misunderstood, was suggested to Darwin by the way in which almost all our varieties of cultivated plants and domestic animals have been obtained from wild forms continually improved for many generations. The method is to breed large quantities, and always preserve or "select" the best in each generation to be the parents of the next. This method, carried on by hundreds of farmers, gardeners, dog, horse, or poultry breeders, and especially by pigeon-fanciers, has resulted in all those useful, beautiful, and even wonderful varieties of fruits, vegetables, and flowers, dray-horses and hunters, greyhounds, spaniels and bull-dogs, cows which give large quantities of the richest milk, and sheep with the greatest quantity and finest quality of wool. All these were produced gradually for the special purposes of mankind; but a similar result has been effected by Nature through rapid increase, great variability, and continual destruction of all the individuals less adapted to the conditions of their special environment, so that only the strongest or the swiftest, the best-concealed or the most wary, the best armed with teeth, horns, hoofs or claws, those who could swim best, or those that protected each other by keeping in flocks or herds, lived the longest and tended to improve still further the next generation. "Survival of the fittest" was suggested by Herbert Spencer as best describing exactly what happens, and it is a most useful descriptive term which should always be kept in mind when discussing or investigating the process by which the infinitely varied and beautiful productions of Nature have been developed. There is really not one single part or organ of any plant or animal that cannot have been derived by means of the fundamental facts of variability and reproduction from some allied plant or animal.

It is interesting here to note that the two *essential factors* of the process of constant adaptation to the environment by great variability and rapid multiplication formed no part of Lamarck's theory, which some people still think to be as good as Darwin's. Equally suggestive is the fact that, while extensive groups of life-phenomena, such as color, weapons, hair, scales, and feathers can hardly be conceived as having been produced or modified by *effort* or by the direct action of the environment, they are yet, every one of them, perfectly explained by the fundamental and necessary processes of variability and survival, acting slowly and continuously, but with intermittent periods of extreme activity at long intervals, on all living things.

One of the weakest and most foolish of all the objections to the Darwinian theory is, that it does not explain *variation*, and is therefore worthless. We might as well say that Newton's discovery of the laws of gravitation was worthless because its *cause* was not and has not yet been discovered, or that the undulatory theory of light and heat is worthless, because the origin of the ether, the thing that *undulates*, is not known. The *beginnings* of things can never be known, and, as Darwin well said, it is foolish to waste time in speculation about them. I think I have shown in my *World of Life* that infinite variability is a basic law of Nature and have suggested its probable purpose. That purpose seems to have been the development of a life-world culminating in Man--a being capable of studying, and enjoying, and to some extent comprehending, the vast universe around him, from the microscopic life in almost every drop of water to the whirling nebulæ of the glittering star-depths extending to almost unimaginable distances around him.

Looking at him thus, man is as much above, and as different from, the beasts that perish as they are above and beyond the inanimate masses of meteoritic matter which, as we now know, occupy the apparently vacant spaces of our solar system, and from which comets and stars are in all probability the aggregations due to the action of the various cosmic forces which everywhere seem capable of producing variety and order out of a more uniform but less orderly chaos.

But besides this lofty intellect, man is gifted with what we term a moral sense: an insistent perception of justice and injustice, of right and wrong, of order and beauty and truth, which as a whole constitute his moral and esthetic nature, the origin and progress of which I have endeavored to throw some light upon in the present volume. The long course of human history leads us to the conclusion that this higher nature of man arose at some far distant epoch, and though it has developed in various directions, does not seem yet to have elevated the whole race much above its earliest condition, at the time when, by the influx of some portion of the spirit of the Deity, man became "a living soul."

We will now consider some of the changes which this higher nature of man has produced in the action of the laws of variation and natural selection. These are very important, and are so little understood that almost all popular writers on the subject of the future of mankind are led into stating as scientific conclusions what are wholly opposed to the actual teaching of evolution.

CHAPTER XIV: SELECTION AS MODIFIED BY MIND

The theory of natural selection as expounded by Darwin was so completely successful in explaining the origin of the almost infinitely varied forms of the organic world, step by step, during the long succession of the geological ages, that it was naturally supposed to be equally applicable to mankind. This was thought to be almost certain when, in his later work, *The Descent of Man*, Darwin proved by a series of converging facts and convincing arguments that the physical structure of man was in all its parts and organs so extremely similar to that of the anthropoid apes as to demonstrate the descent of both from some common ancestor.

So close is this resemblance that every bone and muscle in the human body has its counterpart in that of the apes, the only differences being slight modifications in their shape and position; yet these differences lead to external forms, attitudes, and modes of life so divergent that we can hardly recognize the close affinity that really exists. This affinity is so real and unmistakable that such a great and conservative zoologist as the late Sir Richard Owen declared that to discover and define any important differences between them was the anatomist's difficulty. It was in the dimensions, the shape, and the proportions of the brain that Owen found a sufficient amount of distinctive characters to enable him to place Man in a separate order of mammals--Bimana, or two-handed--while the remainder of the whole monkey tribe--including the apes, baboons, monkeys, and lemurs--formed the order Quadrumana, or four-handed animals. This classification has been rejected by most modern biologists, who consider man to form a distant family only--Hominidæ--of the order Primates, which order includes all four-handed animals as well as man.

But if we recognize the brain as the organ of the mind, and give due weight to the complete distinctness and enormous superiority of the mind of man as compared with that of all other mammals, we shall be inclined to accept Owen's view as the most natural; and this becomes almost certain when we realize the enormous effect his mind has produced, in modifying and almost neutralizing the action of that great law of natural selection which has held supreme sway in every other portion of the organic world.

We have seen in the preceding chapter how every form of organic life during all the vast extent of geological time has been subject to the law of natural selection, which has incessantly moulded their bodily form and structure, external and internal, in strict adaptation to the successive changes of the world around them, while that world was itself hardly, if at all, modified by them. A few isolated cases--such as the formation of islands by the coral-forming zoophytes, or the damming of a few rivers by the rude though very remarkable labors of the beaver--can hardly be considered as forming exceptions to this law.

But so soon as man appeared upon the earth, even in the earliest periods at which we have any proofs of his existence, or in the lowest state of barbarism in which we are now able to study him, we find him able to use and act upon the forces of Nature, and to modify his environment, both inorganic and organic, in ways which formed a completely new departure in the entire organic world.

Among the very rudest of modern savages the wounded or the sick are assisted, at least with food and shelter, and often in other ways, so that they recover under circumstances that to most of the higher animals would be fatal. Neither does less robust health or vigor, or even the loss of a limb or of eyesight, necessarily entail death. The less fit are therefore not eliminated as among all other animals; and we behold, for the first time in the history of the world, the great law of natural selection by the survival only of "the fittest" to some extent neutralized.

But this is only the first and least important of the effects produced by the superior faculties of man. In the whole animal world, as we have seen, every species is preserved in harmony with the slowly changing environment by modifications of its own organs or faculties, thus gradually leading to the production of new species equally adapted to the new environment as its ancestor was before the change occurred.

In the case of man, however, such bodily adaptations were unnecessary, because his greatly superior mind enabled him to meet all such difficulties in a new and different way. As soon as his specially human faculties were developed (and we have as yet no knowledge of him in any earlier condition), he would cease to be influenced by natural selection in his physical form and structure. Looked at as a mere animal he would remain almost stationary, the changes in the surrounding universe ceasing to produce in him that powerful modifying effect which they exercise over all other members of the entire organic world. In order to protect himself from the larger and fiercer of the mammalia he made use of weapons, such as stone-headed clubs, wooden spears, bows and arrows, and various kinds of traps and snares, all of which are exceedingly effective when families or larger groups combine in their use. Against the severity of the seasons he protected himself with a clothing of skins, and with some form of shelter or well-built house, in which he could rest securely at night, free from tempestuous rain or the attacks of wild beasts. By the use of fire he was enabled to render both roots and flesh more palatable and more digestible, thus increasing the variety and abundance of his food far beyond that of any species of the lower animals. Yet further, by the simplest forms of cultivation, he was able to increase the best of the fruits, the roots, the tubers, as well as the more nutritious of the seeds, such as those of rice and maize, of wheat and of barley, thus securing in convenient proximity to his dwelling-place an abundance of food to supply all his wants and render him almost always secure against scarcity or famine or disastrous droughts.

We see, then, that with the advent of Man there had come into existence a being in whom that subtle force we term *mind* became of far more importance than mere bodily structure. Though with a naked and unprotected body, *this* gave him clothing against the varied inclemencies of the seasons. Though unable to compete with the deer in swiftness or with the wild bull in strength, *this* gave him weapons with which to capture or overcome both. Though less capable than most other animals of living on the herbs and the fruits that unaided Nature supplies, this wonderful faculty taught him to govern and direct Nature to his own benefit, and compelled her to produce food for him almost where and when he pleased. From the moment when the first skin was used as a covering, when the first rude spear was formed to assist him in the chase, when fire was first used to cook his food, when the first seed was sown or shoot planted, a grand revolution was effected in Nature--a revolution which in all previous ages of the earth's history had had no parallel. A being had arisen who was no longer subject to bodily change with changes of the physical universe--a being who was in some degree superior to Nature, inasmuch as he knew how to control and regulate her action, and could keep himself in harmony with her, not through any change in his body, but by means of his vast superiority in mind.

The view above expounded of the transference of the action of natural selection from the bodily structure to the mind of early man was my first original modification of that theory, having been communicated to the *Anthropological Review* in 1864. It received the approval both of Darwin himself and of Herbert Spencer, and I am not aware that anyone has shown any flaw in the reasoning by which it is established. It is certainly of high importance, since if true it renders impossible any important change in the external form of mankind, while it serves as an explanation of the complete identity of specific type of the three great races of man--the Caucasian or white, the Mongolian or yellow, and the Negroid or black--in every essential of human form and structure, while in their best examples they approach very nearly to the same ideal of symmetry and of beauty. Yet so little attention has been given to this view that most popular and even some scientific writers take it for granted that no such difference exists between man and the lower animals. They assume that we are destined to have our bodies modified in the remote future in some unknown way, and that the idea that there is anything approaching final perfection in the human form is a mere figment of the imagination.

Others are so imbued with the universality of natural selection as a beneficial law of Nature that they object to our interfering with its action in, as they urge, the elimination of the unfit by disease and death, even when such diseases are caused by the insanitary conditions of our modern cities or the misery and destitution due to our irrational and immoral social system. Such writers entirely ignore the undoubted fact that affection, sympathy, compassion form as essential a part of human nature as do the higher intellectual and moral faculties; that in the very earliest periods of history and among the very lowest of existing savages they are fully manifested, not merely between the members of the same family, but throughout the whole tribe, and also in most cases to every stranger who is not a known or imagined enemy. The earliest book of travels I remember hearing read by my father was that of Mungo Park, one of the first explorers of the Niger. He was once alone and sick there, and some negro women nursed him, fed him, and saved his life; and while lying in their hut he heard them singing about him as the poor white man, of whom they said:--

> "He has no mother to give him milk,
> No wife to grind his corn."

Hospitality is, in fact, one of the most general of all

human virtues, and in some cases is almost a religion. It is an inherent part of what constitutes "human nature," and it is directly antagonistic to the rigid law of natural selection which has universally prevailed throughout the lower animal world. Those who advocate our allowing natural selection to have free play among ourselves, on the ground that we are interfering with Nature, are totally ignorant of what they are talking about. It is Nature herself, untaught unsophisticated *human* nature, which they are seeking to interfere with. They seek to degrade the higher nature to the level of the lower, to bring down Heaven-born humanity, in its essential characteristics only a little lower than the angels, to the infinitely lower level of the beasts that perish.

The conclusion reached in the earlier portion of this volume, that the higher intellectual and moral nature of man has been approximately stationary during the whole period of human history, and that the cause of the phenomenon has been the absence of any selective agency adequate to increase it, renders it necessary to give some further explanation as to the probable or possible origin of this higher nature, and also of that admirable human body which also appears to have reached a condition of permanent stability.

CHAPTER XV: THE LAWS OF HEREDITY AND ENVIRONMENT

In dealing with the great problems of organic development there is probably no department in which so much error and misconception prevails as on the nature and limitations of Heredity. These misconceptions not only pervade most popular writings on the subject of evolution, but even those of men of science and of specialists in biology, and they are the more important and dangerous because their promulgators are able to quote Herbert Spencer, and to a less extent Darwin, as holding similar views.

The subject is of special importance here because it involves the question of whether the effects of the environment, including education and training, are in any degree transmitted from the individuals so modified to their progeny--whether they are or are not cumulative. It is, in fact, the much discussed and vitally important problem of the Heredity of Acquired Characters. The effects of use and disuse, another form of the same general phenomenon, were assumed by Lamarck to be inherited, and a large portion of his theory of evolution rested on this assumption; it seemed so probable, and was apparently supported by so many facts, that Darwin, like most other naturalists at the time, accepted it without any special inquiry, and when he worked out his theory of Pangenesis in order to explain the main facts of heredity, his suppositions were adapted to include such phenomena. Let us then first explain what is meant by the "acquired characters" which it was thought that a true theory of heredity must explain.

As a rule, the great majority of the peculiarities of any species of animal or plant are constantly reproduced in its offspring. The short tail of the wren, the much longer tail of the long-tailed tit, the crest of the crested tit and of innumerable other birds, always when full-grown exhibit the same characters as in their parents. These are said to be *innate* characters. In rare cases, however, offspring are born which differ materially from their parents, as when a white blackbird or a six-toed kitten appears, but these are equally *innate*, and are often strongly inherited. All these are subject to variation, and can therefore be modified by selection, whether natural or artificial, and the effects of such selection in the case of domestic animals is often enormous. Such are the pouters and tumblers among pigeons, the bull-dog and the greyhound, the numerous breeds of poultry, all of which are known to have been produced by artificial selections of favorable variations extending over many centuries; and the characters of these varieties are all strongly inherited.

Characters which are acquired during the life of the individual owing to differences in the use of certain organs or of exposure to light, heat, drought, wind, moisture, etc., are comparatively very slight, and are liable to be so combined with *innate* characters and with the effects of natural or artificial selection, that it is exceedingly difficult to ascertain, without such careful and long-continued experiments as have not yet been made, whether they are in any degree transmissible from parent to offspring, and therefore cumulative.

Almost every individual case of supposed inheritance of such characters, when carefully examined, has been found to be explicable in other ways; but there is a very large amount of *general* evidence, demonstrating that even if a certain small amount of such inheritance exists, it can certainly not be a factor of any importance in the process of organic evolution, all the factors of which must be universally present because the process itself is universal. I will therefore here limit myself to a short enumeration of a few of the very numerous cases in which the continued use of an organ does not strengthen or improve it, but often the reverse, and of others in which it cannot be asserted that the action of the environment can have had any part whatever in the continuous change or specialization of the part or organ. The number, size, form, position, and composition of the teeth of all the mammalia are extremely varied, and throughout the whole class afford the best characters to distinguish family and generic groups; they are therefore of great value in determining the affinities of extinct forms, because the jaws and teeth, especially the latter, are most frequently preserved. But as the permanent teeth are fully formed while buried in the jawbones and covered by the gums, it is quite certain that the special adaptation of the teeth of each species to seize, crush, tear, or grind up its particular food cannot possibly have been produced by the act of feeding, the effect of which is almost always to grind away the teeth and render them less serviceable. Such adaptation could not possibly have been produced by *use* alone, or any other direct action of the environment. Yet, as the adaptation is clear, and often very remarkable, some eminent palæontologists have declared it to be proved that the changes in them were *produced* by the changes in the environment, and that they constitute very strong evidence of the "inheritance of acquired characters"--a statement unsupported by any direct evidence.

The same objection applies to most of the special organs of sense. The internal organ of hearing is a highly complex series of bones and membranes, protected by the outer ear; but it cannot be even imagined to have been gradually developed by the action of the air waves, the vibrations of which it conveys to the brain.

The eye is a still more striking case, as too much use injures or even destroys it; while specialties of vision, as long or short sight, are undoubtedly *innate*, and usually persist throughout life.

So the wonderfully varied bills of birds cannot be conceived as having been modified by use, and are, in fact, unchangeable when once formed. Yet, as they vary largely in every species, they are readily modified, so as to become adapted to new conditions by the "survival of the fittest."

Equally impossible is it to connect any use or disuse, or environmental action, in the production, the gradual development, or complete adaptation to their conditions of life of the outer coverings of almost all living things--the hair of mammalia, the feathers of birds, the scales or horny skins or solid shields of reptiles, the solid shells of molluscs, wonderfully ribbed or spined, whorled, or turreted, and infinitely varied in surface color and markings. Even more conclusive are the facts presented by the vast hosts of the insect world, from the massive armor of the ever-present beetle tribe, more varied in form, structure, ornament, and color than any other comparable group of living things, to the widely different lepidoptera, equalling, or perhaps surpassing, the whole class of birds in their marvellous grace and beauty, yet all utterly beyond any possible direct action of the environment or of use and disuse in their development, and their close adaptation to that environment.

Organic nature is indisputably one and indivisible. It has been developed throughout by means of the fundamental *forces* of life, of growth and reproduction, and the equally fundamental *laws* of variation, heredity, and enormous increase, resulting in a perpetual adaptation in form, structure, color, and habits to the slowly changing environment. These forces and laws are *universal* in their action; they are demonstrably adequate to the production of the whole of the phenomena we are now discussing. We see, then, that over by far the greater part of the whole world of life any modification of external structure, form, or coloring during the life of the individual is impossible, while in the remainder its action, if it exists at all, is of very limited range. No adequate *proof* of the inheritance of the slight changes thus caused has ever yet been given, and it is therefore wholly unnecessary and illogical to assume its existence and to adduce it as having any part in the ever-active and universal process of evolution.

Throughout the whole series of the animal world, and especially in the higher groups which approach nearest to ourselves, mental and physical characters are so inextricably intermixed in their relation to the laws of evolution and heredity, that either of them, studied separately, leads us to the same conclusions. We are not, therefore, surprised to find that breeders of animals of all kinds act upon the principle that all the qualities of the various stocks, whether bodily or mental, are *innate* and have been due to selection; while training, though necessary to bring out the good qualities of the individual, has had no part in the *production* of those qualities. When a horse or dog of good pedigree is accidentally injured so that it cannot be regularly trained, it is still used for breeding purposes without any doubt as to its conveying to its progeny the highest qualities of its parentage.

In the case of the human race, however, many writers thoughtlessly speak of the hereditary effects of strength or skill due to any mechanical work or special art being continued generation after generation in the same family, as among the castes of India. But of any progressive improvement there is no evidence whatever. Those children who had a natural aptitude for the work would, of course, form the successors of their parents, and there is no proof of anything hereditary except as regards this *innate* aptitude.

Many people are alarmed at the statement that the effects of education and training are *not* hereditary, and think that if that were really the case there would be no hope of improvement of the race; but closer consideration will show them that if the results of our education in the widest sense, in the home, in the shop, in the nation, and in the world at large, had really been hereditary, even in the slightest degree, then indeed there would be little hope for humanity; and there is no clearer proof of this than the fact that we have not *all* been made much worse--the wonder being that any fragment of morality, or humanity, or the love of truth or justice for their own sakes still exists among us.

If we glance through the past history of mankind we see

an almost unbroken succession of aggression and combat between the various races, nations, and tribes. We can dimly see that this continual struggle did lead to a rather severe process of selection, as in the lower animal world. It can hardly be doubted that as a result of these struggles the strongest physically, the most ingenious in the use of weapons, and the best organized for war did survive, and that the weaker and lower were either exterminated or kept as slaves by the conquerors. This leads to alternation of success and failure. We see great conquerors and great material civilizations as a result of their accumulations of wealth and of slaves. Then, for a time, luxury and the arts flourished, and with them came rulers who encouraged degradation and vice at home, supported by more and more remote conquests. Then new conquerors arose, often lower in civilization--barbarians, as they were termed--but higher in the simple domestic virtues and a more natural life of productive labor. These again, or some portions of them, rose to luxury and civilization, to lives of gross sensuality and the most cruel despotism, till outraged humanity raised up new conquerors to go over again the old terrible routine.

The periods of culmination of these old civilizations, founded always on conquest, massacre, and slavery, are marked out for us by the ruins of great cities, temples, and palaces, often of wonderful grandeur, and with indications of arts, science, and literature, which still excite our admiration in Egypt and India, Greece and Rome, and thence through the Middle Ages down to our own time. But the inhumanities and horrors of these periods are inconceivable. A gloomy picture of them is given in that powerful book, *The Martyrdom of Man*, by Winwood Reade; and they are summarized in Burns' fine lines:

"Man's inhumanity to man
Makes countless thousands mourn."

Think of the horrors of war in the perpetual wars of those days before the "Red Cross" service did anything to alleviate them. Think of the old castles, many of which had besides the dungeons a salaried torturer and executioner. Think of the systematic tortures of the centuries, of the witchcraft mania and of the Inquisition. Think of the burnings in Smithfield and in every great city of Europe. Think of

"Truth for ever on the scaffold,
Wrong for ever on the throne."

Freedom of speech, even of thought, were everywhere crimes: how, then, did the love of truth survive as an ideal of today? To escape these horrors, the gentle, the good, the learned, and the peaceful had to seek refuge in monasteries and nunneries, while by means of the celibacy of the clergy the Church, as Galton tells us, "by a policy singularly unwise and suicidal, brutalized the breed of our forefathers."

Here was the actual *education* of the world as man rose from barbarism to civilization, and it was accompanied by a certain amount of retrograde selection by the cruel punishments, confinement in dungeons, or torture and death of those who opposed the rulers, and by the survival of the worst tools of the lords and tyrants. Ought we not to be thankful that such education and custom, the varied influences of such an environment, *were not* hereditary? And is not the fact that the whole world has *not* become utterly degraded, and that anything good remains in our cruelly oppressed human nature, an overwhelming proof that such influences *are not* hereditary?

When we remember that many of these degrading laws and customs, oppressions, and punishments have extended down to our own times; that the terrible slave-trade and the equally terrible slavery have only been abolished within the memory of many of us, and that the system of wage-slavery, the distinction of classes, the gross inequality of the law, the overwork of our laboring millions, the immoral luxury and idleness of our upper-class thousands, while far more thousands die annually of want of the bare necessaries of life; that millions have their lives shortened by easily preventable causes, while other millions pass their whole lives in continuous and almost inhuman labor in order to provide means for the enjoyments and pernicious luxuries of the rich--we must be amazed at the fact that there is nevertheless so much real goodness, real humanity, among us as certainly exists, in spite of all the degrading influences that I have been compelled here to enumerate.

To myself, there seems only one explanation of the very remarkable and almost incredible results just stated. It is, that the Divine nature in us--that portion of our higher nature which raises us above the brutes, and the influx of which makes us men--cannot be lost, cannot even be permanently deteriorated by conditions however adverse, by training however senseless and bad. It ever remains in us, the central and essential portion of our *human* nature, ready to respond to every favorable opportunity that arises, to grasp and hold firm every fragment of high thought or noble action that has been brought to its notice, to oppose even to the death every falsehood in teaching, every tyranny in action. The ethics of Plato and of the great moralists of the Ciceronian epoch, together with those of Jesus and of His disciples and followers, kept alive the sacred flame of pure humanity, and their preservation constitutes perhaps the greatest service the monastic system rendered to the human race. This service is finely expressed by an almost unknown poet, J. H. Dell, in the prefatory to his volume, *The Dawning Grey*. Never has our indebtedness to the classical writers been more powerfully insisted on than in the following lines:--

"Hear ye not the measured footfalls echoing solemn and
sublime,

From the groves of Academus down the avenues of Time;
See'st thou not the giant figures of the Sages of the Past,

Through the darken'd long perspective on the living
foreground cast;

Feel'st thou not the thrilling rhythm of the grand old Grecian
line,

Pulsing to the march of Progress, cadencing her hymn divine,

All the forces of the present by the subtle sparks controlled,

Of the quickening Grecian fire, of the mighty Lights of old?

"Through the dark and desolation of the centuries between,

Still 'The Porch's' glories glimmer, still 'The Garden's' wreaths
are green.

Still the Zeno, still the Plato, still the Pyrrho points the page,

Still the Philip fears the pebble--still Melitus dreads the Sage,

Still the Dionysius trembles at the stylus of the age.

Still the dauntless ranks of Freedom kindle to Tyrtæus' song;

Still they bear aloft the symbol--bear the glorious torch along."

If the Christian Church had done nothing for us but
preserve in its monasteries and abbeys the finest examples of
classic literature that have come down to us, and given us
those glories of Gothic architecture which seem to express
in stone the grandeur and sublimity, the peacefulness and
the beauty of a pure religion, it would, notwithstanding its

many defects, its cruelty and oppression, its opposition to the study of nature and to freedom of thought, have fully justified its existence as helping us to realize whatever more advanced and purer civilization the immediate future may have in store for us.

Some Light on the Problem of Evil

Before passing on to another branch of my subject I feel it necessary to make a few suggestions in reply to the objection that will certainly and very properly be made, as to why, if our higher human nature is in its essence Divine, it has suffered such long and terrible eclipses--why has the lower so often and for so long prevailed over the higher? This is, of course, one of the many forms of the old problem of the origin of evil, which is no doubt insoluble by us. But as it is a fairly well-defined and limited portion of that problem it may be possible to obtain some idea of a possible solution, and as such a one has occurred to myself during the composition of the present volume, I will give it as briefly as possible in the hope that it may interest some of my readers.

In my recent works, *Man's Place in the Universe* and *The World of Life*, the conclusion was forced upon me that the scheme of the development of the universe of stars and nebulæ with which we are acquainted, and especially of our sun and solar system, was such as to furnish the exact conditions on our earth, and there only, which should allow of the origin and evolution of the organic world culminating in man. Yet further, that the conditions should be such as to produce the maximum of diversity both of inorganic and organic products useful to man, and such as would aid in the development of the greatest possible diversity of character and especially of his higher mental and moral nature. What I have here termed the Divine influx, which at some definite epoch in his evolution at once raised man above the rest of the animals, creating as it were a new being with a continuous spiritual existence in a world or worlds where eternal progress was possible for him. To prepare him for this progress with ever-increasing diversity, faculties of enormous range were required, and these needed development in every direction which earthly conditions rendered possible. In order that this extreme diversity of character should be brought about, a great space of time, as measured by successive generations, was necessary, though utterly insignificant as compared with the preceding duration of organic life on the earth, and still more insignificant as compared with the spirit-life to succeed it. It is for this purpose, perhaps, that languages become so rapidly diverse and mutually unintelligible after a moderate period of isolation, binding together small or moderate communities in distinct tribes or nations, which each develop in their own way under the influence of special physical surroundings and originate peculiarities of habits, customs, and modes of thought. Antagonisms soon arise between adjacent tribes, leading each to protect itself against others by means of chiefs and some quasi-military combinations. This requires organization and foresight, and after a time the most powerful conquers the weaker, they intermingle, and still greater diversity arises. By this constant struggle the less advanced suffer most, and the race as a whole takes a step forward in the march of civilization.

We see the best example of this mode of progress by antagonism in the small States of Ancient Greece, where each little kingdom developed its peculiar form of art, of government, and of civilization, which it transferred to all parts of Europe; and after two thousand years of degradation by Roman and Turkish conquest, its language still remains but little altered, while its ancient literature and art are still unsurpassed. In like manner Rome brought law, literature, and military discipline to an equally high level; and it too sank into a state of ruin and degradation, while its literature and its law continued to illuminate the civilized world during its long struggle towards freedom. Wherever conditions were favorable to progress in art or science, *time* was needed for its full growth and development, while perpetual war necessitated organization and training against conquest or destruction. Even the cruelties and massacres by despotic rulers excited at last the uprising of the oppressed, and so developed the nobler attributes of patriotism, courage, and love of freedom. In the very worst of times there was an undercurrent of peaceful labor, art, and learning, slowly moulding nations towards a higher state of civilization.

The point of view now suggested will perhaps be rendered somewhat more intelligible if we apply it to the nineteenth century, of which I have written in such condemnatory terms. The preceding eighteenth century was undoubtedly a somewhat stationary epoch, of a rather commonplace character alike in literature, in art, in science, and in social life. Its vices also were low, its government bad, its system of punishments cruel, and its recognition of slavery degrading. It was a kind of "dark age" between the literary and national brilliance of the Elizabethan age and the wonderful scientific and industrial advance of the Victorian age.

But this latter period was also a period of a great uprising of the specially human virtues of justice, of pity, of the love of freedom, and of the importance of education; and though the rapid increase of wealth through the utilization of natural forces led to all the evils due to the unchecked growth of individual riches and power, yet these very evils in all their intensity and horror were perhaps necessary to excite in a sufficient number of minds the determination to get rid of them. Time was also required for the workers to learn their own power, and, very gradually, to learn how to use it. The rick-burning and machine-breaking of the early part of the century have been succeeded by combination and strikes; step by step political power has been gained by the masses; but only now, in the twentieth century, are they beginning to learn how to use their strength in an effective manner. There are, however, indications that the whole march of progress has been dangerously rapid, and it *might* have been safer if the great increases of knowledge and the vast accumulations of wealth had been spread over two centuries instead of one. In that case our higher nature might have been able to keep pace with the growing evils of superfluous wealth and increasing luxury, and it might have been possible to put a check upon them before they had attained the full power for evil they now possess.

Nevertheless, the omens for the future are good. The great body of the more intelligent workers are determined to have JUSTICE. They insist upon the abolition of monopolies of the forces of nature, and upon the gradual admission of all to *equal opportunities* for labor by free access to their native soil. Thus may be initiated the birth of a new era of peaceful reform and moral advancement.

Note.--As many of my readers may not understand the allusions in the second verse of Mr. Dell's poem (p-130), I append the explanation:

"The Porch," the place where the Stoic philosophers taught--

The Painted Porch in Athens.

"The Garden," scene of Plato's and Socrates' teaching. Zeno was the founder of the Stoic philosophy.

Pyrrho was the founder of the Sceptic school.

Philip of Macedon lost an eye at the siege of Methone by a slinger's pebble.

Melitus was one of the disputants with Socrates, and was always vanquished by him.

Dionysius, the Tyrant of Syracuse, was also a Poet and was a candidate for the prize at the Olympic games, but was conquered and therefore feared the more skilful "stylus" (pen) of the victors. Tyrtæus, a lame schoolmaster of Athens, inspired the Lacedæmonians by his patriotic war-songs, and thus contributed largely to their victories.

CHAPTER XVI: MORAL PROGRESS THROUGH A NEW FORM OF SELECTION

Many readers, and some writers of books on organic evolution, seem quite unaware that Darwin established two modes of selection, both alike "natural" but acting in different ways and producing somewhat different results. He termed the second mode "sexual selection," and in his *Origin of Species* he briefly describes it as consisting in the fighting of males for the possession of females, which undoubtedly occurs in numbers of the higher vertebrates and also in insects.

But he also includes under sexual selection another mode of rivalry by the display of the special male ornaments of many birds, and the choice of the more ornamental by the females. To this latter phase he devotes nearly half his volume on *The Descent of Man*, and on *Selection in Relation to Sex*. Selection by the fighting of males has led to the development of the stag's antlers, the boar's tusks, and the lion's mane serving as a shield. These combats rarely lead to the death of the vanquished, but to a larger number of offspring for the victor, and this leads to the improvement of the race by keeping up its strength, vigor, and fighting power.

The other form of selection, by the display of ornaments by male birds and the supposed continuous development of those ornaments by the appreciative choice of the females, I believe to be imaginary. I have discussed this subject in many of my books, and my views are now generally adopted by evolutionists. The fact that the colors of male insects, especially butterflies, are almost exactly parallel to those of birds, first led me to this conclusion, because we can hardly suppose insects to be endowed with any æsthetic sense, even if they really *see* color at all, which, in my last book, I have given strong reasons for doubting.

But in the human race the conditions are altogether different; for while, as I have shown in Chapter XIV, the kind of natural selection which through all the ages had moulded the infinitely varied animal forms into harmony with their environment ceased to act upon man's body and only for a limited time upon his lower mental faculties, sexual selection tended to act, if at all prejudicially, through polygamy, prostitution, and slavery, though it possesses the potentiality of acting in the future so as to ensure intellectual and moral progress, and thus elevate the race to whatever degree of civilization and well-being it is capable of reaching in earth-life.

Eugenics, or Race Improvement through Marriage

The total cessation of the action of natural selection as a cause of improvement in our race, either physical or mental, led to the proposal of the late Sir F. Galton to establish a new science, which he termed Eugenics. A society has been formed, and much is being written about checking degeneration and elevating the race to a higher level by its means. Sir F. Galton's own proposals were limited to giving prizes or endowments for the marriage of persons of high character, both physical, mental, and moral, to be determined by some form of inquiry or examination. This may, perhaps, not do much harm, but it would certainly do very little good. Its range of action would be extremely limited, and so far as it induced any couples to marry each other for the pecuniary reward, it would be absolutely immoral in its nature, and probably result in no perceptible improvement of the race.

But there is great danger in such a process of artificial selection by experts, who would certainly soon adopt methods very different from those of the founder. We have already had proposals made for the "segregation of the feeble-minded," while the "sterilization of the unfit" and of some classes of criminals is already being discussed. This might soon be extended to the destruction of deformed infants, as was actually proposed by the late Grant Allen; while Mr. Hiram M. Stanley, in a work on *Our Civilization and the*

Marriage Problem, proposed more far-reaching measures. He says: "The drunkard, the criminal, the diseased, the morally weak, should never come into society. Not reform, but prevention, should be the cry." And he hints at the methods he would adopt, in the following passages: "In the true golden age, which lies not behind but before us, the privilege of parentage will be esteemed an honor for the comparatively few, and no child will be born who is not only sound in body and mind, but also above the average as to natural ability and moral force." And he concludes: "The most important matter in society, the inherent quality of the members of which it is composed, should be regulated by trained specialists."

Of course, our modern eugenists will disclaim any wish to adopt such measures as are here hinted at, which are in every way dangerous and detestable. But I protest strenuously against any direct interference with the freedom of marriage, which, as I shall show, is not only totally unnecessary, but would be a much greater source of danger to morals and to the well-being of humanity than the mere temporary evils it seeks to cure. I trust that all my readers will oppose any *legislation* on this subject by a chance body of elected persons who are totally unfitted to deal with far less complex problems than this one, and as to which they are sure to bungle disastrously.

It is in the highest degree presumptuous and irrational

to attempt to deal by compulsory enactments with the most vital and most sacred of all human relations, regardless of the fact that our present phase of social development is not only extremely imperfect but, as I have already shown, vicious and rotten at the core. How can it be possible to determine by legislation those relations of the sexes which shall be best alike for individuals and for the race, in a society in which a large proportion of our women are forced to work long hours daily for the barest subsistence, with an almost total absence of the rational pleasures of life, for the want of which thousands are driven into wholly uncongenial marriages in order to secure some amount of personal independence or physical well-being?

Let anyone consider, on the one hand, the lives of the wealthy as portrayed in the society newspapers of the day, with their endless round of pleasure and luxury, their almost inconceivable wastefulness and extravagance, indicated by the cost of female dress and the fact of a thousand pounds or more being expended on the flowers for a single entertainment. On the other hand, let him contemplate the awful lives of millions of workers, so miserably paid and with such uncertainty of work that many thousands of the women and young girls are driven on the streets as the only means of breaking the monotony of their unceasing labor and obtaining some taste of the enjoyments of life at whatever cost; and then ask himself if the legislature which cannot remedy *this* state of things should venture to meddle with the great problems of marriage and the sanctities of family life. Is it not a hideous mockery that the successive governments which for forty years have seen the people they profess to govern so driven to despair by the vile conditions of their existence that in an ever larger and larger proportion they seek death by suicide as their only means of escape-- that governments which have done nothing to put an end to this continuous horror of starvation and suicide should be thought capable of remedying some of its more terrible *results*, while leaving its *causes* absolutely untouched?

It is my firm conviction, for reasons I shall give farther on, that, when we have cleansed the Augean stable of our present social organization, and have made such arrangements that *all* shall contribute their share either of physical or mental labor, and that every one shall obtain the full and equal reward for their work, the future progress of the race will be rendered certain by the fuller development of its higher nature acted on by a special form of selection which will then come into play.

When men and women are, for the first time in the course of civilization, alike free to follow their best impulses; when idleness and vicious or hurtful luxury on the one hand, oppressive labor and the dread of starvation on the other, are alike unknown; when *all* receive the best and broadest education that the state of civilization and knowledge will admit; when the standard of public opinion is set by the wisest and the best among us, and that standard is systematically inculcated on the young; then we shall find that a system of *truly natural* selection will come spontaneously into action which will steadily tend to eliminate the lower, the less developed, or in any way defective types of men, and will thus continuously raise the physical, moral, and intellectual standard of the race. The exact mode in which this selection will operate will now be briefly explained.

Free Selection in Marriage

It will be generally admitted that, although many women now remain unmarried from necessity rather than from choice, there are always considerable numbers who feel no strong impulse to marriage, and accept husbands to secure subsistence and a home of their own rather than from personal affection or strong sexual emotion. In a state of society in which all women were economically independent, were all fully occupied with public duties and social or intellectual pleasures, and had nothing to gain by marriage as regards material well-being or social position, it is highly probable that the numbers of the unmarried from choice would increase. It would probably come to be considered a degradation for any woman to marry a man whom she could not love and esteem, and this reason would tend at least to delay marriage till a worthy and sympathetic partner was encountered.

In man, on the other hand, the passion of love is more general and usually stronger; and in such a society as here postulated there would be no way of gratifying this passion but by marriage. Every woman, therefore, would be likely to receive offers, and a powerful selective agency would rest with the female sex. Under the system of education and public opinion here supposed, there can be little doubt how this selection would be exercised. The idle or the utterly selfish would be almost universally rejected; the chronically diseased or the weak in intellect would also usually remain unmarried, at least till an advanced period of life, while those who showed any tendency to insanity or exhibited any congenital deformity would also be rejected by the younger women, because it would be considered an offence against society to be the means of perpetuating any such diseases or imperfections.

We must also take account of a special factor, hitherto almost unnoticed, which would tend to intensify the selection thus exercised. It is a fact well known to statisticians that, although females are in excess in almost all civilized populations, yet this is not due to a law of Nature; for with us, and I believe in all parts of the Continent, more males than females are born to an amount of about 3 1/2 to 4 per cent. But between the ages of five and thirty-five there were, in 1910, 4.225 deaths of males from accident or violence and only 1.300 of females, showing an excess of male deaths of 2.925 in one year; and for many years the numbers of this class of deaths have not varied much, the excess of preventable deaths of males at those ages being very nearly 3,000 annually. This excess is no doubt due to boys and young men being more exposed, both in play and work, to various kinds of accidents than are women, and this brings about the constant excess of females in what may be termed normal civilized populations.

In 1901 it was about a million; while fifty years earlier, when the population was about half, it was only 359,000, or considerably less than half the present proportion. This is what we should expect from the constant increase of accidents and of emigration, the effects of both of which fall most upon males.

It appears, therefore, that the larger number of women in our population today is not a natural phenomenon, but is almost wholly the result of our own man-made social environment. When the lives of *all* our citizens are accounted of equal value to the community, irrespective of class or of wealth, a much smaller number will be allowed to suffer from such preventable causes; while, as our colonies fill up with a normal population, and the enormous areas of uncultivated or half-cultivated land at home are thrown open to our own people on the most favorable terms, the great tide of emigration will be diminished and will then cease to affect the proportion of the sexes. The result of these various causes, now all tending to increase the numbers of the female population, will, in a rational and just system of society, of which we may hope soon to see the commencement, act in a contrary direction, and will in a few generations bring the sexes first to an equality, and later on to a majority of males.

There are some, no doubt, who will object that, even when women have a free choice, owing to improved economic conditions, they will not choose wisely so as to advance the race. But no one has the right to make such a statement without adducing very strong evidence in support of it. We have for generations degraded women in every possible way; but we now know that such degradation is not hereditary, and therefore not permanent. The great philosopher and seer, Swedenborg, declared that whereas men loved justice, wisdom, and power for their own sakes, women loved them as seen in the characters of men. It is generally admitted that there is truth in this observation; but there is surely still more truth in the converse, that they do not admire those men who are palpably unjust, stupid, or weak, and still less those who are distorted, diseased, or grossly vicious, though under present conditions they are often driven to marry them. It may be taken as certain, therefore, that when women are economically and socially free to choose, numbers of the worst men among all classes who now readily obtain wives *will be almost universally rejected.*

Now, this mode of improvement by elimination of the less desirable has many advantages over that of securing early marriages of the more admired; for what we most require is to improve the *average* of our population by rejecting its lower types rather than by raising the advanced types a little higher. Great and good men are always produced in sufficient numbers and have always been so produced in every phase of civilization. We do not need more of these so much as we want a diminution of the weaker and less advanced types. This weeding-out process has been the method of *natural selection*, by which the whole of the glorious vegetable and animal kingdoms have been developed and advanced. The survival of the fittest is really the extinction of the unfit; and it is the one brilliant ray of hope for humanity that, just as we advance in the reform of our present cruel and disastrous social system, we shall set free a power of selection in marriage that will steadily and certainly improve the character, as well as the strength and the beauty, of our race.

Social Reform and Over-population

One of the most general and apparently the strongest of

the objections to any thorough schemes of social reform, and especially to those that will abolish want and the constant dread of starvation is that, in any society in which this is done early marriages will be much more numerous; there will be no prudential checks to large families; and in a few generations, as Malthus argued, populations will increase beyond the means of subsistence. Then will commence a continual decrease of well-being, culminating in universal poverty, worse than any that now exists, because it will be universal. The following quotation from an eminent American writer shows that this fear has really been felt:

"If it be true that reason must direct the course of human evolution, and if it be also true that selection of the fittest is the only method available for that purpose; then, if we are to have any race-improvement at all, the dreadful law of *destruction of the weak and helpless* must, with Spartan firmness, be carried out voluntarily and deliberately. Against such a course all that is best in us revolts."* [*Professor Joseph Le Conte, in *The Monist*, Vol. I., .]

A more recent writer, Dr. W. M. Flinders Petrie, the well-known Egyptian explorer, has put forward similar views in a tentative manner, but clearly showing what he thinks our present state of society requires. Of the compensation to workmen for accident he says:

"The immediate effect upon character is to save the careless, thoughtless and incompetent from the results of their faults; this at once reduces largely the weeding and educational effects of the bad qualities."

And of old-age pensions his concluding remark is:

"Nature knows of no right to maintenance, but only the necessity of getting rid of these who need it by mending or ending them."

Again, as to the huge waste of infant life now going on, which he admits is preventable and might be saved, he remarks:

"We must agree that it would be of the lower, or lowest type of careless, thriftless, dirty, and incapable families that the increase would be obtained. Is it worth while to dilute our increase of population by 10 per cent. more of the more inferior kind?"

And he concludes thus:

"This movement is doing away with one of the few remains of natural weeding out of the unfit that our civilization has left us. And it will certainly cause more misery than happiness in the course of a century."* [**Janus in Modern Life*. By W. M. Flinders Petrie, D.C.L., F.R.S.]

The whole book is full of such statements as the above, for which neither facts nor arguments are given. It is assumed throughout that the failures in our modern society are so through their own fault--they are "wastrels"--and deserve neither pity nor help. He knows nothing apparently of Dr. Barnardo's work in rescuing these "wastrel" children from the gutter and the workhouse, treating them well and kindly, training them in work, and sending many thousands to Canada. A record of their subsequent life was kept, and it was found that very few failed to do well, while a very large majority became valuable citizens in their new home. On the whole, they were in no way inferior to the average of emigrants who go at their own expense, and who are admitted to be among the best of our workers.

None of the writers of the class here quoted seem to have made themselves acquainted with the researches of Herbert Spencer, Sir F. Galton, and others as to the natural laws which determine the rate of increase of population when those laws are allowed to operate freely under rational and moral social conditions. A short statement of these laws will therefore be given.

In a remarkable essay, first published in 1852, H. Spencer, with his usual philosophical insight, examined the facts of reproduction and population throughout the whole of the animal kingdom, and showed that the duration of the individual life and the increase of the race varied inversely, those groups which have the simplest organization and the shortest lives producing the greatest number of offspring; in other terms, individuation and reproduction are antagonistic. But individuation depends almost entirely on the development and specialization of the nervous system, through which alone all advance in instinct, emotion, and intellect is rendered possible. The actual rate of increase in man has been determined by the necessities of the savage state, in which, as in most species of mammals, it is usually what is just required to maintain a limited average population. But with a true advance in civilization the average duration of life increases, and the possible increase of population under favorable conditions becomes very great because fertility is greater than is needed under the new conditions. At present, however, no general advance in intellectuality has taken place; but that the facts do accord with the theory is indicated by the common observation that highly intellectual parents do not have large families, while the most rapid increase occurs in those classes which are engaged in healthy manual labor.

But a law founded on such a broad physiological basis of observation is sure to continue in action, and we may therefore feel certain that as the intellectual level of the whole race is raised by general culture and physical health, the law of diminishing fertility will act, and will tend in the remote future to bring about an exact balance between the rate of increase and that of mortality.

A more immediate and effective check to rapid increase of population will, however, be brought about by the social reforms already suggested. When poverty is abolished and neither economic nor social advantages will be gained by early marriage, there can be no doubt it will be generally deferred to a later age. Still more effective will be the extension of the period of education or training for the whole population for several years longer than at present, together with the growth of public opinion against all marriages between persons who have not yet begun the serious work of life. It would also be an essential part of education to inculcate the delay of marriage till every opportunity has been afforded both of the parties concerned of becoming thoroughly acquainted with each other before undertaking so serious a responsibility as marriage usually involves.

The effect of even a few years' delay of marriage on population is very considerable. Sir F. Galton has shown from the best statistics available that if we compare women married at twenty with those at twenty-nine, the comparative fertility is as 8 to 5. But this does not represent the whole effect on increase of population. When marriage is delayed, the time between successive generations is correspondingly increased; and yet another effect in the same direction is produced by the fact that the greater the average age of marriage the fewer generations are alive at the same time, and it is the combined effect of these three factors that determines the actual increase of the population due to this cause.

Sir F. Galton gives a remarkable table showing this combined result of these causes. He finds that if one hundred mothers and their daughters in each successive generation marry at twenty, there will be an increase of such mothers in each successive generation of 1.15. If, however, they marry at twenty-nine, each successive generation of mothers diminishes in the proportion of 0.85. If this goes on for 108 years, the hundred mothers who marry at twenty have increased to 175, and in 216 years to 299; while those who marry at twenty-nine will have decreased to 61 and 38 respectively. It is therefore shown that under present social conditions the age of marriage necessary to preserve a stationary population will be somewhere between twenty and twenty-nine. The above figures are, however, founded on special cases, and the actual facts are so complicated by the number of childless marriages, the rate of infantile mortality and other causes, that they must be taken only as establishing a *law* of rather rapid decrease of fertility with each year's addition to the average age of marriage of the mother.

I have now, I venture to hope, established two important principles in relation to human progress. In the first place, I have shown that modern ideas as to the necessity of dealing *directly* with some of our glaring social evils, such as race degeneration and the various forms of sexual immorality, are fundamentally wrong and are doomed to failure so long as their fundamental causes--widespread poverty, destitution, and starvation--are not greatly diminished and ultimately abolished. I have proved that human nature is *not* in itself such a complete failure as our modem eugenists seem to suppose, but that it is influenced by fundamental laws which under reasonably just and equal *economic conditions* will automatically abolish all these evils.

In the second place, I have shown that the dread of over-population as the result of the abolition of poverty is wholly and utterly fallacious--a mere bugbear created by ignorance of natural laws and of presumption in thinking that we can cure social evils while leaving the man-made causes which produce them unaltered. The three great natural laws which all our would-be reformers ignore are:

(1) That a very moderate advance in the average age of marriage--which would certainly result from a truly rational system of education combined with economic equality--necessarily diminishes the rate of increase of the population.

(2) That every approach to educational and economic equality by effecting a large saving of the lives of males who now die from preventable causes, combined with the fact that male births *exceed* those of females, would so diminish the number of the latter that they would soon become less instead of, as now, more than that of males: that this would give them an *effective choice* in marriage which they do not now possess, together with the power of delay which for many reasons large numbers of them would exercise.

(3) The law of diminishing fertility with increase of brain-work through education and training would further tend to the diminution of fertility.

These three natural causes all *tend* in one direction--the equality of births with deaths, while their action would be so readily modified by public opinion as to obviate all danger of either increase or decrease beyond what was necessary for the well-being of each community, nation, or race.

The Future Status of Woman

The foregoing statement of the effect of established natural laws, if allowed free play under rational conditions of civilization, clearly indicates that the position of woman in the not distant future will be far higher and more important than any which has been claimed for or by her in the past.

While she will be conceded full political and social rights on an equality with man, she will be placed in a position of responsibility and power which will render her his superior, since the future moral progress of the race will so largely depend upon her free choice in marriage. As time goes on, and she acquires more and more economic independence, *that* alone will give her an effective choice which she has never had before. But this choice will be further strengthened by the fact that, with ever-increasing approach to equality of opportunity for every child born in our country, that terrible excess of male deaths, in boyhood and early manhood especially due to various preventable causes, will disappear, and change the present majority of women to a majority of men. This will lead to a greater rivalry for wives, and will give to women the power of rejecting all the lower types of character among their suitors.

It will be their special duty so to mould public opinion, through home training and social influence, as to render the women of the future the regenerators of the entire human race. We hope and believe that they will be fully equal to the high and responsible position which, in accordance with natural laws, they will be called upon to fulfil.

The certainty that this powerful selective agency will come into existence just in proportion as we reform our existing social system by the abolition of poverty and the establishment of full equality of opportunity in education and economic position, demonstrates that Nature--or the Universal Mind--has not failed or bungled our world so completely as to require the weak and ignorant efforts of the eugenists to set it right, while leaving the great fundamental causes of all existing social evils absolutely untouched. Let them devote all their energies to purifying this whitened sepulchre of destitution and ignorance, and the beneficent laws of human nature will themselves bring about the physical, intellectual, and moral advancement of our race.

CHAPTER XVII: HOW TO INITIATE AN ERA OF MORAL PROGRESS

In Chapters VIII to XII of this volume I have given in briefest outline a summary of the growth during the nineteenth century of the actual social environment in the midst of which we live.

We see a continuous advance of man's power to utilize the forces of Nature, to an extent which surpasses everything he had been able to do during all the preceding centuries of his recorded history.

We also see that the result of this vast economic revolution has been almost wholly evil.

We see that this hundredfold increase of wealth, amply sufficient to provide necessaries, comforts, and all beneficial refinement, and luxuries for our whole population, has been distributed with such gross injustice that the actual condition of those who produce all this wealth has become worse and worse, no efficient arrangements having been made that from the overflowing abundance produced *all* should receive the mere essentials of a healthy and happy existence.

We have seen huge cities grow up, every one of them with their overcrowded, insanitary slums, where men, women, and children die prematurely as surely as though a body of secret poisoners were constantly at work to destroy them.

We see thousands of girls compelled by starvation to work in such an empoisoned environment as to produce horribly painful and disfiguring disease, which is often fatal in early youth, or in what ought to have been, and what might have been, the period of maximum enjoyment of their womanhood. And to this very day no efficient steps have been taken to abolish these conditions.

We see millions still struggling in vain for a sufficiency of the bare necessaries of life (which in their misery is all they ask), often culminating in actual starvation, or in suicide, to which they are driven by the dread of starvation. Yet our Governments, selected from among the most educated, the most talented, the wealthiest of the country, with absolute power to make what laws and regulations they please, and an overflowing fund of accumulated wealth to draw upon, do nothing, although more people die annually of want than are killed in a great war, and more children than could be slaughtered by many Herods.

And while all this goes on in the depths, where--

"Pale anguish keeps the heavy gate,
And the Warder is Despair"--

a little higher up, among the middle-men distributors of

the necessaries and luxuries of life, bribery, adulteration, and various forms of petty dishonesty are rampant.

And higher yet, among the great Capitalists, the merchant Princes, the Captains of industry, we find hard taskmasters who drive down wages below the level of bare subsistence, and who support a more gigantic and widespread system of gambling than the world has ever seen.

And, finally, our administration of what we call "Justice" (and of which we are so proud because our judges cannot be bribed) is utterly *unjust*, because it is based on a system of money fees at every step; because it is so cumbrous and full of technicalities as to need the employment of attorneys and counsel at great cost, and because all petty offences are punishable by fine *or* imprisonment, which makes poverty itself a crime while it allows those with money to go practically free.

Taking account of these various groups of undoubted facts, many of which are so gross, so terrible, that they cannot be overstated, it is not too much to say that our whole system of society is rotten from top to bottom, and the Social Environment as a whole, in relation to our possibilities and our claims, is the worst that the world has ever seen.

Such are the evil products of the social environment we have ourselves created in the course of a single century. We have seen it going from bad to worse, and have applied petty remedies here and there during the whole period; but the evils have continued to increase. It has now become clear to the more intelligent of the workers that if we wish to improve it--if we wish to prevent it from getting even worse than it is--we must deal with the root-causes of the evil and, so far as possible, *reverse the conditions which are so demonstrably bad, such hideous failures.* And fortunately, this is by no means so difficult as it may seem to be, because a large body of our thinkers and a considerable number of our workers see clearly what these root-causes are, and, less clearly, how to remedy them. They will, however, give their energetic support to any Government that devotes itself to the task of remedying them. The following are my own views as to how the problem must be attacked in order to solve it thoroughly and permanently.

The Root-cause and the Remedy

If we review with care the long train of social evils which have grown up during the nineteenth century, we shall find that every one of them, however diverse in their nature and results, is due to the same general cause, which may be defined or stated in a variety of different ways:

(1) They are due, broadly and generally, to our living under a system of universal *competition* for the means of existence, the remedy for which is equally universal *co-*

operation.

(2) It may be also defined as a system of *economic antagonism*, as of enemies, the remedy being a system of *economic brotherhood*, as of a great family, or of friends.

(3) Our system is also one of *monopoly* by a few of all the means of existence: the land, without access to which no life is possible; and capital, or the results of stored-up labor, which is now in the possession of a limited number of capitalists and therefore is also a monopoly. The remedy is freedom of access to land and capital for all.

(4) Also, it may be defined as *social injustice*, inasmuch as the *few* in each generation are allowed to inherit the stored-up wealth of all preceding generations, while the *many* inherit nothing. The remedy is to adopt the principle of equality of opportunity for all, or of universal *inheritance by the State in trust for the whole community.*

These four statements of the existing *causes* of all our social evils cannot, I believe, be controverted, and the *remedies* for them may be condensed into one general proposition: that it is the first duty (in importance) of a civilized Government to organize the labor of the whole community for the equal good of all; but it is also their first duty (in time) to take immediate steps to abolish *death by starvation and by preventable disease* due to insanitary dwellings and dangerous employments, while carefully elaborating the *permanent* remedy for want in the midst of wealth.

I myself have pointed out how these two ends may be best achieved, and hope to elaborate them. In the meantime, I call attention to Mr. Standish O'Grady's letter, "To the Leaders of Labor," in *The New Age* of November 21, 1912, in which, after referring to the very natural dread by the rich of any such radical reorganization of society, as leading to their own financial ruin (which it certainly need not do), he makes the following suggestive statement, with which I hope all my readers will agree:

"But what they fail to perceive is, that, in a world like this, made by infinite goodness and wisdom, Right is always the great stand-by for men and for Nations, and for the rich as well as for the poor; and that Wrong, sooner or later, ends in misery and destruction."

That is sound moral teaching. We have been doing the wrong for the past century, and we have reaped, and are reaping, "misery and destruction." It is time that we changed our methods, which are all (as I think I have sufficiently pointed out) fundamentally wrong, radically unjust, wholly immoral.

We have ourselves created an immoral or unmoral social environment. To undo its inevitable results we must reverse our course. We must see that *all* our economic legislation, *all* our social reforms, are in the very opposite direction to those hitherto adopted, and that they tend in the direction of one or other of the four fundamental remedies I have suggested.

In this way only can we hope to change our existing immoral environment into a moral one, and initiate a new era of Moral Progress.

In Chapters XIII to XVI I have shown that the well-established laws of evolution as they really apply to mankind are all favorable to the advance of true civilization and of morality. Our existing competitive and antagonistic social system alone neutralizes their beneficent operation. That system must therefore be radically changed into one of brotherly co-operation and co-ordination for the equal good of all. To succeed we must make this principle our guide and our polestar in all social legislation.